CONCEPTS OF MODERN PHYSICS

The Haifa Lectures

CONCEPTS OF MODERN PHYSICS

The Haifa Lectures

Mendel Sachs
The State University of New York, Buffalo, USA

ICP
Imperial College Press

Published by

Imperial College Press
57 Shelton Street
Covent Garden
London WC2H 9HE

Distributed by

World Scientific Publishing Co. Pte. Ltd.
5 Toh Tuck Link, Singapore 596224
USA office: 27 Warren Street, Suite 401-402, Hackensack, NJ 07601
UK office: 57 Shelton Street, Covent Garden, London WC2H 9HE

British Library Cataloguing-in-Publication Data
A catalogue record for this book is available from the British Library.

Cover image: "Haifa" by Michael Cohn.

CONCEPTS OF MODERN PHYSICS

ISBN-13 978-1-86094-821-3
ISBN-10 1-86094-821-9
ISBN-13 978-1-86094-822-0 (pbk)
ISBN-10 1-86094-822-7 (pbk)

Typeset by Stallion Press
Email: enquiries@stallionpress.com

Printed in Singapore by World Scientific Printers (S) Pte Ltd

To the Memory of

Abraham Kaplan and Nathan Rosen

Philosopher and Physicist par excellence

PREFACE

This book is based on a lecture course that I taught in Israel, entitled 'Concepts of Modern Physics.' It was given in the spring term, 2006, in the Philosophy Department, University of Haifa.

The main emphasis of this course is the *concepts* of modern physics: the quantum theory and the theory of relativity, and their relation to problems of modern astrophysics and cosmology. The discussion is largely from the point of view of the ideas of modern physics, rather than from their mathematical expressions. [*The reader may skip some of the unavoidable mathematical terms, especially in Lecture VI on the quantum theory, without losing too much of the thread of ideas of the entire book.*]

As a professional physicist, I learned in the early stages of my career that physics without philosophy does not fulfill the goal of science. It does fulfill its obligation of providing empirical descriptions of the ways of nature. This is the *descriptive stage* of science. But it is essential that one must then proceed to the *explanatory stage*, including some of the philosophical understandings. Einstein made the following poignant comment: "Physics without Philosophy would be blind; Philosophy without Physics would be lame."

Some who attended my Haifa lectures asked this question: Is not a *complete* description of a phenomenon in nature a sufficient explanation of it? My answer is No! The goal of science — the truths about the natural world — comes in the two stages. The first stage is at the *descriptive* level. It is based on experimentation. Once this is achieved, one must then use rational analyses as well as human intuition to proceed to the second stage — the stage of *explanation*. The explanation is in terms of underlying principles. With logical and mathematical analyses, one may then deduce *from these alleged first principles* the features of the derived facts of natural phenomena. If these are in agreement with the empirical facts, one may then claim (at least contigently!) to have achieved an increase in our comprehension of the real world.

This course consists of ten lectures: I. Philosophy of Science, II. Classical Precursors for the Concepts of Modern Physics, III. Nineteenth Century Physics: Atomism and Continuity, IV. Early Anomalies and Elementary Particles, V. From the Old Quantum Theory to Quantum Mechanics, VI. Quantum Mechanics: Heisenberg's Matrix Mechanics and the Copenhagen School, VII. Concepts of the Theory of Relativity, VIII. From Special to General Relativity, IX. The Universe, X. Conflicts in the Foundations of the Quantum and Relativity Theories.

The idea of this course of lectures is to discuss modern physics primarily from a critical point of view — *thus the view of philosophy* — as well as the view of the history of science.

The ideas expressed in these lectures evoked controversial discussions about the concepts of modern physics. Out of these controversies, I believe that: (1) There developed an incitement for the participants in the course to think for themselves about the ideas of science, (2) the idea was conveyed that controversy is essential for progress in science (and in any other fields of ideas that search for the truth), (3) a belief that we can never reach a *complete* understanding of the real world. Thus, physics remains a never-ending adventure into unknown territory. We have indeed achieved more comprehension about the real world at this stage of our history than was achieved in past times. But we should also know that our comprehension of the world, so far achieved, is infinitesimal in comparison with what there is to be understood!

Mendel Sachs
Buffalo, New York, September, 2006

ACKNOWLEDGMENTS

I thank Robert R. Sachs for his excellent and critical reading of these lectures. His philosophical insights were invaluable. I also thank the students and faculty who attended these lectures in the Department of Philosophy, University of Haifa, Israel, for their pertinent questions and comments.

CONTENTS

Lecture I

PHILOSOPHY OF SCIENCE

Introduction

In the following lectures, I wish to discuss in a mainly non-mathematical way, the concepts of modern physics. The word "concepts" refers to the *ideas* that underlie modern physics.

It is true that for the training of professional physicists, to make progress in the exploration of the truths of nature one must learn the mathematical language to facilitate an expression of the laws of nature. Plato advised that: for the students to become educated, they must first master the field of mathematics. Once this is done, he continued, they are ready to study the field of physics. After this, they are finally ready to study philosophy.

It is interesting that about 15 centuries after Plato, in the Middle Ages, the Jewish philosopher/theologian, Rambam (known in the western world as Maimonides) added to Plato's advice. He wrote a book aimed at the 'true' student, called The Guide for the Perplexed *(University of Chicago, 1963, transl. S. Pines). In this treatise he advised the student that after following Plato's advice, in mastering mathematics, physics and philosophy, he was ready to study the spiritual side of the real world — to attempt to understand the Biblical Scriptures in terms of first principles.*

I do not generally agree that an education necessitates Plato's first requirement — a broad knowledge of how to facilitate and to understand the intricacies of mathematics. Galileo correctly commented that *the book of nature is written in the language of mathematics.* However, according to Plato's own teaching, a language, *per se,* is not the set of concepts that it is meant to express. In this course, we will stress the concepts of physics rather than its mathematical language.

The adjective, "modern," in the title of this course, refers to 20th century physics, from the concepts that underlie elementary particle physics — the physics of the smallest domain — to those of cosmology — the physics of the universe as a whole.

It is significant that the 20th century has witnessed the confluence of two scientific revolutions, rather than one at a time. These are the theory of relativity, led primarily by Albert Einstein, and the quantum theory, led primarily by Niels Bohr. What was fascinating about two rather than one revolution in science was that, on the one hand, each of these theories is incompatible with the other, both mathematically and conceptually, on foundational terms. But on the other hand, each of these theories requires the other for its completion. This is the major problem of modern physics that will be emphasized throughout this course of lectures.[1]

Because of their conceptual and mathematical incompatibility, it follows that if one of these two dichotomous theories of matter is true, the other must be false, as foundational truths of nature, even though each has been empirically successful in its own domain of nature. A way out of this dilemma would be to fully reject the underlying concepts of one of these theories while generalizing the other theory so as to recover the empirically successful features of the rejected theory, *as a mathematical approximation.* [This relies on the well-known *principle of correspondence* to tell us how to approach the formal expression of an earlier rejected theory — where it works! — from a later accepted theory in science.] I will argue, toward the end of the course, that, in my judgment, it is the theory of general relativity that will survive to *explain* the basic nature of matter in all domains — from elementary particle physics to cosmology. But the main thrust of the course will be to *explain* the basic concepts of both the quantum and relativity theories. The readers may then make up their own minds as to which of these theories is more true to nature than the other, in all domains.

At this juncture, I wish to qualify my earlier use of the term "scientific revolution." I do not believe that the history of science is indeed characterized by 'revolutions' — a total rejection of earlier concepts and replacing them with entirely new concepts to explain natural phenomena.[2] I do believe, based on my study of the history of science, that science progresses in an 'evolutionary' manner. Here, there are 'threads of truth' that persist from one period of 'normal science' to another. Indeed, it is the persistence of these threads of truth that gives us confidence that some of the past ideas play an important role in the shaping of present day ideas as well as the future of science.

It is for this reason that I will begin this course (in Lecture II) with ideas from the past that serve as important precursors for present day ideas and the physics of the future. But before we start it is important to discuss some of the ideas of the philosophy of science.

Philosophy of Science

The first question we must answer is: What is the meaning of "physics"? The word was introduced by Aristotle in ancient Greece. It signifies the 'science of inanimate matter.'[3] Of course, there are other sciences — *psychology* is the science of human consciousness, *anthropology* is the science of human cultural systems, *biology* is the science of life systems, etc. But it is thought that physics is the queen of the sciences because it is framed in foundational terms, starting with first principles, while the other sciences, at this stage of our understanding, are more phenomenological — more descriptive than explanatory. Hopefully, some of these latter sciences will proceed to foundational levels at a later stage.

Then what is the meaning of the word "science"? It is the pursuit toward the fundamental *truths* of the natural world. These are expressed in terms of the *laws of nature*. That is to say, the fundamental goals of science are the 'truths' of the natural world. These truths, in turn, are at the *explanatory level* of our understanding.

They follow from the *descriptive level* of scientific knowledge. Though some scientists believe that the descriptive level of science is all that there is to know, that is, they believe that scientists should only ask 'what' questions, I believe that the explanatory level that follows the descriptive level is the actual goal of science — the answers to the 'why' questions. It is the approach that provides the *understanding* of the natural world that we seek. The former philosophical (epistemological) view is that of *positivism* while the latter view is that of *realism*. The latter view was that of Einstein, when he said: "the most incomprehensible thing to me is that we can comprehend anything at all about the real world."[4]

Truth

Our primary goal in science, then, is to acquire the truth about the natural world. It is important to be cognizant, in this regard, that there are different kinds of truth, defined in different contexts.

As I see it, there are three types of truth that we signify in science. The first is 'analytic truth.' This is an irrefutable type of truth — it is a *necessary truth*. All mathematical truths are of this sort. For example, the statement 2 + 3 = 5, is a necessary truth. Defining the integers 2 and 3 as intervals along a linear scale, and the logical definitions of + and =, the conclusion 5 is a necessary, irrefutable truth. It is not tied to nature — it comes from the thinking of the human mind that leads to the invention of the rules of a logical system.

The second type of truth is *scientific truth*. It is, in principle, a *refutable truth* — it is 'contingent.' The conclusion of a scientific truth starts with a *guess* about a law of nature to underlie some physical phenomenon. One arrives at this law from hints received from our perceptions of the phenomenon itself as well as from our intuition. The alleged law of nature then leads, by logical deduction, to particulars that are the predictions of empirical facts of nature. If these predictions agree with the empirical facts, then we may say that the original law that led to these results is *scientifically true*. But this is not a necessary truth. It is only as true as the premises that led to the alleged law of nature. Such a law may be confirmed in terms of different empirical facts predicted by this law for a very long time, perhaps centuries, but it may then be refuted by new empirical evidence that is not predicted by this law of nature, or a demonstration of logical inconsistency in its expression. An example is Newton's second law of motion, $F = ma$. It was empirically verified from the 17th century until the 20th century. Then, Einstein's relativistic mechanics refuted and replaced it. Einstein had then refuted Newton's alleged law of motion. It is important, then, that a scientific truth, tied to nature, is in principle refutable, while an analytic truth, tied to logic, is in principle not refutable. Thus, they are in different contexts. One may then not answer a scientific question with an analytic argument, or answer an analytic question with a scientific argument — they would be non-sequiturs. For example, Pythagoras, in ancient Greece, claimed to come to scientific conclusions about the structure of the universe from mathematical (analytic) arguments.[5] This is a false claim, as it is a non-sequitur.

A third type of truth is a *religious truth*. It is a truth that is based on faith. This truth is irrefutable as is an analytic truth, but it is not subject to the rules of an invented logical system, as is mathematics. A belief in God is an example of a religious truth. Another example is the scientists' belief that for every effect in the world, there is an underlying cause. (*This is a law of total causation.*) Indeed, it is the *raison d'etre* of the scientist to pursue the cause–effect relation that is a law of nature. The pursuit is based on his or her *faith* in the law of total causation. Because this type of truth is in a different context than the analytic or scientific truths, one cannot prove a scientific statement (or an analytic statement) with a religious argument. It would be a non-sequitur. Because scientific truth is in a different context than religious truth, a scientist may have faith in the existence of God (or not have such faith) without giving up his or her integrity as a scientist!

An example that demonstrates this fallacy is in the subject of cosmology (as we will discuss more fully in Lecture IX). It was discovered in the 1920s by E. Hubble, that the galaxies of the universe are moving away from each other at a particular rate, in accordance with the 'Hubble law,' $v = Hr$, where v is the speed of one galaxy relative to another and r is their mutual separation.[6] Thus, as r increases, v increases linearly — i.e. the galaxies are accelerating relative to each other's positions, linearly. That is, if, for example, the distance between galaxies doubles, so would their relative speed of separation double. This is called the "expansion of the universe." [*What is meant by this expansion is not that the universe as a whole is expanding into empty space — because there is no empty space! The universe, by definition, is all that there is! What is meant by the expansion of the universe is that the density of the matter of the universe, at any location within the universe, relative to any observer, is decreasing with respect to any observer's time measure.*]

Thus if we extrapolate backwards in time, we come to greater and greater densities of the matter of the universe. In the limit, we come to a maximum density and a maximum instability. At that point in time, there is predicted to be a gigantic explosion — the 'big bang' — when all of the matter of the universe starts its expansion phase. From the present day astrophysical measurement of the expansion, the Hubble constant H may be determined and the time when the 'big bang' happened may be estimated. It is found to be between 12 and 15 billion years ago.

The *scientific question* then arises: How did all of the matter of the universe get into the state of maximum density and instability in the first place? Some (including a number of astrophysicists and physicists) have answered: The 'big bang' occurred when God created the universe! But this *religious answer* to a scientific question is a non-sequitur! Indeed, a bona fide scientific question requires a bona fide scientific answer!

The only answer that I see is that before the matter of the universe reached the critical state of maximum density and instability, it was contracting (imploding) from a less dense state toward this more dense state. And before the contraction, the matter of the universe was expanding, and so on *ad infinitum*. This is the *oscillating universe cosmology*. We will discuss it in more detail toward the end of this course of lectures. Suffice it to say at this point that the theory of general relativity allows for this resolution because the parts of this theory that relate to the force of matter on matter (the affine connection terms) are not 'positive-definite.' This implies that gravitational forces entail both attractive and repulsive forces. Under the conditions of sufficiently high density of matter and relative speeds, the repulsive force dominates. This leads to the expanding universe — that matter repels other matter. But when the matter becomes sufficiently rarefied, the attractive force dominates the repulsive force and the expansion changes to a contraction.

This view then predicts, from General Relativity, that the universe is a closed system, undergoing a continual cyclic process of expansion and contraction. The 'big bang' we talk about, then, about 12 to 15 billion years ago, is only the beginning of this particular cycle. One might then ask: when was the *actual* 'beginning' of the universe — the 'beginning' of all of the cycles? (This is the 'beraisheet' — the opening word of the Bible.) This is not a scientific question! It is a religious question. It is not definable in terms of our concept of 'time.' The idea of 'timelessness' is reflected in one of the Kabbalistic names for God — *'haya-hoveh-yeeheeyeh* (*was-is-will be*).

We see here how important it is to clearly distinguish between scientific truth and religious truth. It is only the former that we focus on in this course of lectures. [*It is the recognition of this difference between scientific truth and religious truth that led the courts in the US to the ruling that the religious assertion, called "intelligent design," may not be taught in the schools as though it were a scientific* explanation *for the creation of the universe — that it would be a violation of the Constitution's separation between church and state.*]

Significant Questions in Science

Finally, on the subject of the philosophy of science, it is noteworthy that the theoretical scientists spend most of their time on either answering questions or asking questions. The former is tied mostly to questions of phenomenology (*description* — the 'what' questions) and the latter is tied mostly to questions of foundation (*explanation* — the 'why' questions).

The difficult thing about asking a question in science is to know in advance if this would be a significant question — that is, a question whose answer could add to our comprehension of the natural world. For it is possible that a scientist may spend most of his or her professional life, as well as the lives of admiring colleagues and students, in trying to answer an insignificant question! Thus it would be useful if there would be some criteria that could help the scientist to decide beforehand if a question in science is indeed significant. One criterion that I have proposed is that if a question leads to *more questions*, it is more likely significant than if it leads to no more questions. The reason is my belief, based on the history of science is that the total understanding of any phenomenon in nature is unbounded — it is infinite in extent. Thus if there would be no more questions about it, the phenomenon would appear to be totally understood — the investigator would claim to have infinite knowledge about it! Since human beings are finite, this would be impossible (i.e. human beings cannot be omniscient)!

I believe that it is for this reason that in the 16th century Galileo said: "there is not a single effect in nature, even the least that exists, such that the most ingenious theorists can arrive at a complete understanding of it."[7] This idea stands against the current claims that physicists have discovered a 'theory of everything!'[8]

In the next lecture, we will present some of the ideas of classical physics from the 16th century onward that were important precursors for the ideas of physics of the 20th century — the quantum and relativity theories.

Lecture II

CLASSICAL PRECURSORS FOR THE CONCEPTS OF MODERN PHYSICS

Galileo Galilei

The Thought Experiment

We start with the researches of Galileo, who lived in the 16th–17th centuries. He is sometimes known as the 'father of modern physics.' It is because he taught us a method for approaching scientific truth that has been followed to this day. It is based on an interweaving of theoretical and experimental physics. We start with the 'thought experiment.' This is an assumed *ideal* experiment in which assumed conditions are not met in the real world, such as the motion of a body on a frictionless plane. This logical experiment then yields conclusions that may be tested in real experiments. If these experiments give results that are in agreement with the conclusions of the thought experiment, then we may say that the original assumptions of the thought experiment are *scientifically true*.

Galileo's Principle of Inertia

An interesting example of the thought experiment is Galileo's deduction of his *Principle of Inertia*. This is the idea that if a material body moves at a constant speed on a horizontal plane, in a straight line, it will maintain this state of motion forever, unless it is made to alter it by some external influence.[9] That is, constant motion of a body is one of its natural attributes (in contradiction with the idea of Aristotle, who believed that rest is the natural state of a material body).

The thought experiment that led Galileo to this conclusion is as follows: Consider a material body at the top of a frictionless inclined plane, at the vertical height h. He then argued that the force that would cause the *free fall* of the body (if the inclined plane were not in the way) is directed toward the center of the earth. The original height h of the body, at the top of the inclined plane, corresponds to the state of 'rest' of the body. After the body is released, it slides down the frictionless plane with ever increasing speed, reaching a maximum speed at the bottom. Suppose it then faces a second inclined plane. It would then proceed to move up this plane. When would it stop? He concluded (intuitively!) that at a continuously slower pace, it must stop when it reaches the same vertical height h it had when it started to move on the first downward plane. If the angle of inclination of the (second) upward slope is less than the angle of inclination of the downward slope, the body would have to go further along the upward plane to reach the vertical height h, than the distance it traveled down the first plane from this height. It would then take more time to come to rest on the upward slope, than it did to travel the length of the downward slope. Thus, if one should decrease the angle of inclination of the second slope to be horizontal, i.e. at 0 degrees, after moving down from the first inclined path, it would then be moving on the horizontal plane and it would never come to rest — because it would never reach the vertical height h. The material body would then move forever at the speed v (that it acquired at the bottom of the first slope). Thus, Galileo concluded his *Principle of Inertia* in this way — *a body moves naturally in a straight line, at a constant speed, forever, if no external forces would act on it.*

Of course, this is an *ideal* experiment because there is no such thing as a frictionless plane. In the real case, the body would continually lose energy because of the frictional forces exerted on it by the horizontal plane, until it would stop. But the *Principle of*

Inertia, as stated, is a scientifically true assertion. It was stated a generation later by Newton, who called it his 'first law of motion.'

Laws of Motion in Two- and Three-Dimensional Space

Galileo discovered that the motion of a body in two or three dimensions may be represented as a superposition of motions in each of the dimensions, separately. It is the way that he discovered the path of projectile motion. The vertical motion is governed by his law of gravity $g = constant$, where g is the downward acceleration due to gravity, while the superposed horizontal motion is governed by his principle of inertia. This superposition of motions then yields the parabolic trajectory of the projectile. *Later on, it was found that such a superposition of motions in the different directions may be expressed in a vectorial form.*[10]

Galileo's Principle of Relativity

A generation before Galileo, Copernicus discovered that he could not explain the motions of the stars in a single night and a day, that he observed if he assumed that the earth is at rest at the center of the universe, and all of the heavenly bodies, including the Sun, revolve about it. He then postulated that the Sun is at the center of the universe and all heavenly bodies, including the Earth, revolve about it. This is Copernicus' heliocentric model of the universe — that the Sun is at the absolute center of the universe. His assertion that the 'Earth moves' was in contradiction with the teachings of Aristotle and with the Church!

The Copernican concept was a difficult one for human beings (and the Church) to accept in his day. They were taught that the universe was created for mankind, who were placed at its absolute center, on Earth. This pre-Copernican idea has been revived in modern times with the assertion of the *anthropic principle* — the idea that the entire universe was created to accommodate the human race![11]

It was in the generation after Copernicus that Galileo asserted the *Principle of Relativity*. He agreed with Copernicus that 'the Earth moves,' but he asserted further that 'motion' *per se*, is *subjective*.

That is to say, the motion of matter is not one of its *objective* features. With this Galilean view, it is just as *scientifically true* to say that, from the Earth's perspective (i.e. frame of reference), the sun revolves about it, as it is to say that from the Sun's perspective, the Earth revolves about it! That is to say, Galileo saw that 'motion' is not an intrinsic, objective property of matter. He then concluded that, in contrast with Copernicus' heliocentric model of the universe, where it is postulated that the Sun is at the center of the universe, *there is no center of the universe*! [*It is often asserted by contemporary physicists that it was discovered in the 15th century that the Earth revolves about the Sun,* and not vice versa. *This statement is not extended to say what Galileo discovered — that motion is subjective and thus that it is as equally true to say that the Sun revolves about the Earth as it is to say that the Earth revolves about the Sun*!]

All that is required in Galileo's view is that the law of nature (in this case a law of gravity) that governs the motion of matter is independent of the perspective from which it is described, in one way or another. This is *Galileo's principle of relativity.*[12] It is an important precursor for *Einstein's principle of relativity* — the primary axiom that underlies the theory of relativity of modern times. A major change in going from Galileo's principle of relativity to that of Einstein is the following: In the former, the spatial coordinates alone are relative to the reference frame in which the laws of nature is expressed, while in the latter the temporal as well as the spatial coordinates are relative to the reference frame.

Threads of Truth – Astronomy

One of Galileo's important contributions was his law of gravity, $g =$ *constant.*[13] That is, the acceleration due to gravity g for a freely falling object, along a radial direction of the Earth toward its center, is independent of the weight or size of the body. Of course, in the real world heavier weights are seen to fall more rapidly than lighter weights. But this is due to the different air resistance to these falling weights. Galileo did an experiment to prove this. He placed a heavy and a light weight on a platform in an evacuated jar. Releasing them simultaneously, he saw that they landed at the bottom at the same time, proving the constancy of g (= 980 cm/sec^2) for any matter, heavy or light, large or small.

Galileo's law of gravity is a good approximation for Newton's law of universal gravitation near the Earth's surface. With the empirical equivalence of gravitational mass (defined in Newton's law of universal gravitation), and inertial mass (defined in Newton's second of motion), it turned out that in Newton's theory, near the Earth's surface $g = GM/R^2 = 980$ cm/sec^2, where G is the universal gravitational constant, M is the mass of the earth and R is the radius of the Earth.

Three hundred years after Newton's discovery, Einstein's theory of general relativity superseded it. This will be discussed later on in the course.

Galileo was the first astronomer to use the telescope to study the heavens. He saw that the moon is not a gas, as Aristotle thought. Rather he discovered that it is a solid mass with craters on its surface. He saw that Jupiter has many moons revolving about it. He (mistakenly) thought that the *Milky Way* — our galaxy — is the entire universe. There were no observations that refuted this conclusion. Two centuries later it was discovered that the *Milky Way* is only one of an infinitude of galaxies of the universe. It was found in the centuries after Galileo that our *Milky Way* has a neighboring galaxy, Andromeda, that forms a binary with our galaxy. [*There is a legend that when Galileo had his telescope focused on Jupiter, he asked a colleague to look at its moons. After the colleague looked, he was upset because what he saw was not what the Church believed. He declared that Galileo was trying to trick him — in setting up pictures of these moons inside of his telescope! Later on, when his colleague was on his deathbed, Galileo whispered in his ear: "You are a good man and will most likely soon go to heaven. On your way up, you will pass Jupiter. When you do, take a look at the vicinity of Jupiter and you will see its moons*!]

Rene Descartes[14]

Rene Descartes was a contemporary of Galileo. It was his idea that the essences of matter are extension and motion. The extension of matter is in space. Since space is continuous, he argued that matter must be continuous, rather than atomistic. On the other hand, he argued that the expansion or contraction of a piece of metal under changes of temperature implies that its constituent atoms are moving away from or toward each other. With this conclusion, he took the atomistic view of matter. But in the final analysis, Descartes did adhere to the continuous view. He saw the fundamental aspects of matter as vortices in a continuous field. Further, he believed that 'motion' is an intrinsic (objective) feature of matter, rather than the subjective feature that Galileo discussed.

These conflicting views of matter and its motion in Descartes are an important precursor for the 20th century dichotomy of wave and particle as foundational elements of matter that arise in the modern quantum theory. This will be addressed in more detail later on in the course.

One problem that concerned Descartes is the dichotomy between *mind* and *matter*. At first, he believed that the fundamental laws of mind and matter are independent, yet they seem to be coupled in some way. For example, I *see* a rose (matter), I *think* about picking it up (mind), I pick it up and put it close to my nose (matter), I sniff it (matter), and I feel good because of its pleasant aroma (mind). Thus, mind and matter are seemingly coupled. Yet, Descartes felt that they are still uncoupled, dynamically. In his view, it is not unlike moving trains on parallel tracks. The trains do not cause each others motions, but there is a correspondence between their motions and positions.

Later on in his career, Descartes seemed to revert to the idea that the laws of mind and matter are unified. He thought that perhaps the pineal gland of the human body is the source of this coupling (since in his day there was no knowledge of the purpose of this gland).

Descartes' initial studies of the science of human consciousness led some to consider him as the first psychologist.

Baruch Spinoza[15]

Spinoza was another of the great contemporaries of Galileo. His method of arriving at the truth was to use the logic of Euclid and the assertion of *self-evident* facts. He strongly disagreed with the separation of mind and matter in Descartes (or with a *fundamental separation* of observer and observed, or of subject and object). These different components of the whole, such as observer and observed, are to Spinoza only the modes of a continuum. His was a view of *holism* — the idea that there are no singular, separable entities in the universe. Rather, everything is a manifestation (a mode) of a continuous whole. It is an idea that carried over to the continuous field concept in Einstein's theory of general relativity in the 20th century, as a foundational theory of matter.

For example, there is symmetry in Spinoza's view of the cause–effect relations. From the perspective of one component (its frame of reference), one entity (mode) of the whole is the source of the cause of an effect on the other. But from a different perspective, from the frame of reference of the entity that was formerly thought of as the affected one; it is instead the cause and the first entity becomes the one affected.

This view brings in an interesting interpretation of the 'time' parameter in the cause–effect relation. For if the cause is earlier and the effect is later, then one may correlate a time change with the evolution from cause to effect. But when the cause and effect change places, the time must then reverse. Thus, time, *per se*, is not an absolute change. It is only a parameter that is used to represent one's environment in terms of cause–effect relations, from one frame of reference or another. This idea is a very important precursor for a subjective view of time as a relative measure, in the theory of relativity that emerged with Einstein's theory in the 20th century.

Isaac Newton

Isaac Newton was born in the same year (1642) that Galileo died. He is known for his discovery of laws of motion, his law of universal gravitation and his discoveries in the field of optics. He is also known as a great mathematician. He discovered calculus; it was useful for the description of variable motion in the laws of nature.

Newton's Three Laws of Motion[16]

Newton's first two laws of motion were derivative from *Galileo's principle of inertia*: If a material body is unimpeded by any external influence, it will move at a constant speed in a straight line, forever. That is, a constant speed (or rest, as a special case) is a natural feature of matter. This statement is also called *Newton's First Law of Motion.*

Thus, there can be no acceleration of a material body without an external force acting on it. *Newton's Second Law of Motion* states that *if an external force F acts on a material body, then it causes it to accelerate at a rate a, in a linear relation, $F = ma$, where m is the constant of proportionality between F and a.* m is a measure of the intrinsic inertia of a body — it is its resistance to a change of state of rest or constant speed. It is called the 'inertial mass' of the body.

Newton's Third Law of Motion does not come from Galileo. It is the assertion that *for any force exerted by a body A on another body B, there is an equal magnitude force in the opposite direction exerted by B on A.* Thus, A–B is a *closed system.* It is easy to see that that the implication of Newton's third law of motion is that, fundamentally, any material system is *closed.* Further, though Newton did not discuss this mathematical implication of his third law (that appears in the theory of relativity of the present period), any mathematical equation that describes a closed system is *nonlinear.*[17] Newton's third law of motion plays a very important role as a precursor for Einstein's theory of general relativity, 300 years later. It is one of the 'threads of truth' that persist throughout the history of science.

Newton's Law of Universal Gravitation

Newton derived his law of universal gravitation to explain the phenomenon of gravity, and to explain Galileo's finding that the acceleration of *any* body in 'free fall' near the Earth's surface is the constant $g = 980$ cm/sec^2.

To start, Newton used Kepler's empirical law that for any orbiting body, $r^3\omega^2 = K$, where r is the distance from the orbiting body to the center of rotation (such as the distance from an orbiting planet to the Sun), ω is the angular frequency of the orbiting body (in radians/sec) and K, Kepler's constant, depends on the parameters of the body at the center of rotation (e.g. those of the Sun in the case of planetary motion). Newton applied Kepler's law to the motion of the moon, orbiting the Earth. He saw that the centripetal force that keeps the moon in its orbit, due to the gravitational force of the earth, is F(gravity) = F_c(centripetal force) = $m_{\text{moon}}a_c \equiv m_{\text{moon}}r\omega^2$, where $a_c = r\omega^2$ is the centripetal acceleration of the orbiting moon. According to Kepler's result for any orbiting body, this gives: $F(\text{gravity}) = m_{\text{moon}}K/r^2$.

For the moon's orbital motion, Newton then found the relation for the force of gravity (that keeps the moon from veering off of its orbit in a straight line path) is $F(\text{gravity}) = F_c = Gm_{\text{moon}}m_{\text{earth}}/r^2$, where $K = Gm_{\text{earth}}$ is Kepler's constant. [*In this formula, m_{earth} is called the "gravitational mass" of the orbiting body — the Earth in this case. It was found empirically to be equal to the inertial mass of the orbiting body.*]

Thus Newton found that the force of gravity is inversely proportional to the square of the distance between the interacting bodies.

Applied to the case of a body falling freely to the Earth near its surface, the force of gravity is (using Newton's second law of motion and his law of gravitation), $mg = GmM/r^2$, where $r = R + h$, R is the radius of the Earth and h is the height of the body with mass m, above the surface of the earth. Thus, Newton derived the constant $g = GM/r^2$. With $r \approx R$, the known parameters of the Earth, M (the mass of the Earth) and R (the radius of the Earth), he then found that $g = 980$ cm/sec^2, in agreement with Galileo's measurements. Thus Newton claimed to *explain* Galileo's law of gravity.

Newton then applied his finding to all gravitationally interacting bodies. He applied his calculus to finding a solution for his equation of the gravitational force: $ma \equiv m(d^2r(t))/dt^2) = -GmM/r(t)^2$, (the minus sign denotes an attractive force). He found that the solution $r(t)$ traces an elliptical path of the orbit of a planet, with the Sun's center of mass at one of the focal points of this ellipse — in agreement with Kepler's observation in regard to the orbit of the planet Mars. [*Newton said that if he had not been standing on the shoulders of two giants, Galileo and Kepler, he would not have been able to see as far as he did.*]

Newton's law of universal gravitation was then applied successfully to other cases. For example, the orbits of all of the planets of our solar system, the tides (caused by the moon's gravitational pull on the oceans of Earth) the trajectory of Halley's Comet (as it was seen by the astronomer Halley in Newton's day). Newton's law of universal gravitation was then billed as a triumph of science. It was claimed that it should apply to all of the material bodies of the universe. His theory predicted that all bodies must rotate about other bodies of the universe in cyclical orbits, as the planets of our solar system rotate about the sun. The universe was then seen as a giant clock, full of spinning wheels — millions of millions of bodies rotating about other bodies.

It should be noted that the prediction of these orbital paths of all bodies follows from the mathematical separation of the functions of spatial coordinates from the functions of temporal coordinates in the mathematical structure of Newton's theory of the gravitational force. In this way, Newton predicted *stationary orbits* for all bodies of the universe. This is in contrast with Einstein's relativity theory, where the time and space coordinates do not objectively separate. In this case, there are no stationary orbits. The observation of the non-stationary orbit of Mercury about the Sun was then a confirmation of a prediction of Einstein's theory of general relativity and a refutation of Newton's prediction. A second important confirmation is the 20th century observation (by Hubble) of the 'expansion' of the universe, as we will discuss toward the end of this course.

Newton's Concepts

Newton objected to the feature of his law of gravity that is 'action at a distance.' This is the result that gravitationally interacting bodies depend on their mutual spatial separation but not on the time. Thus, if a stellar body, millions of light-years from us should suddenly explode out of existence, we should respond *immediately*. It seemed wrong to Newton. But (perhaps out of frustration) he dismissed the objection with the claim "I do not form hypotheses!" Still, he had the following criticism of this concept: "Action at a distance through a vacuum, without mediation of anything else by and through which their action and force may be conveyed from one to another, is to me so great an absurdity, that I believe no man who has in philosophical matters competent faculty of thinking, can ever fall into it." [*Both Faraday, of the 19th century, and Einstein, of the 20th century, agreed with Newton's assessment of the error in the concept of 'action at a distance.' But in contrast with Newton, they tried to do something about replacing it with a paradigm change in physical thought*!]

Newton's view of the infinitude of the universe is atomistic. It is made up of individual, separable 'things' that interact (or not) with each other. This model is a precursor for the particle view of matter of quantum mechanics (except for its determinism versus the indeterminism of quantum mechanics). But this is in contrast with Einstein's theory of relativity wherein the world of matter is *holistic* — there are no separate, individual, material particles in the first place. The universe is a continuum whereby the 'things' of it are the modes of this continuum, similar to the ripples of a pond.[18]

Newton's Optics

As a generalist, Newton believed that if the material world is atomistic, so must 'light' be atomistic. He explained the dispersion of white light (as the light coming from the Sun) as it propagates through a dispersive medium, such as the droplets of water vapor after a rainstorm, or through a prism, as a combination of different colored particles of light — red, orange, yellow, green, blue, indigo, violet — each having the same speed in a vacuum but different speeds in dispersive media. Thus they emerge from a dispersive medium at different times, separating out, as in the view of the rainbow.

With this model he also explained in a mechanical fashion the reflection of light (from a perfect mirror) and refraction of light, i.e. the bending of a light beam as it propagates from one medium to a denser medium). But he could not explain the diffraction of light. He looked for this effect, but he did not see it in his own experimentation. [Leonardo da Vinci saw the diffraction of light effect 200 years before Newton's time!] An example of the diffraction effect is that of a monochromatic beam of light moving through a screen with holes in it (or past a hair, as in Newton's own experiment) and landing on a second screen with interference maxima and minima. (Newton did not have enough resolution in his experiment to see this effect.) The interference phenomenon is characteristic of waves, not particles! In Newton's day, Robert Hooke and Christiaan Huygens believed in the wave theory of light and, with experimentation, they showed its validity, including the refraction effect and that of the diffraction of light. In the final analysis, it was the wave theory of light that won over the particle theory. The Maxwell theory of electromagnetism, discovered in the 19th century, showed that light is a manifestation of the electromagnetic field, and all of the known features of optics follow from the 'radiation solutions' of Maxwell's equations, representing light as a wave phenomenon. In the 20th century it appeared that units of light (photons) seemed to be particles in some experimentation and waves in other experimentation. This led, in modern times, to the concept of *wave–particle dualism*, which we will discuss later on in our analysis of the quantum theory.

In the next lecture, we will discuss the ideas of 19th century physics, yielding the particle and continuous field views as fundamental, under separate types of experimentation. It is pertinent that the science of the 19th century served as an important bridge between the classical period and that of the revolutionary movements of 20th century physics.

Lecture III

NINETEENTH CENTURY PHYSICS: ATOMISM AND CONTINUITY

The *threads of truth* that carried on through the 19th century, since the earlier classical period of Galileo and Newton, focused on two major trends. One was the description of matter explained with a model of atomism, as studied initially in terms of the gaseous state of matter. This led, in turn, to the laws of gases, thermodynamics and the conservation of energy.

The second trend was toward a view opposite to that of atomism — the continuous field concept as a fundamental notion. It was applied initially to the nature of electricity and magnetism, and thence to an understanding of optics and other radiation phenomena. The atomistic view is based on the idea that all matter is composed of singular, separable, independent *things*, that may or may not interact with each other. The continuous field concept is one of *holism*, wherein the underlying basis of matter is in terms of continuous fields that have as manifestations modes that give the impression of things, but are in reality all correlated through one continuous entity. [*One might see this view as akin to that of Spinoza, in the 16th century — which, in the 20th century, Einstein acknowledged to be very influential in his understanding of the nature of matter. The latter holistic view did not come to full fruition in physics until the discovery of Einstein's theory of general relativity.*]

The Ideal Gas Law

Matter occurs in the universe in three distinct states: gas, solid and liquid, depending on the physical conditions of its environment. At the beginning of the 19th century a great deal of attention was focused on the gaseous state of matter. Measurements were made of the pressure P and the temperature τ of a gas, confined to a volume V. It was found that, to a first approximation, the relationship between these parameters of the gas is: $PV = R(\tau + 273)$, where R is the *universal gas constant*, and τ is the temperature in *degrees Celsius*, as read on a thermometer (that is calibrated so that $\tau = 0$ *degrees Celsius* is the temperature where water freezes and $\tau = 100$ *degrees Celsius* is the temperature where water boils). Changing the temperature scale to $T = (\tau + 273)$ *degrees absolute*, the ideal gas law is: $PV = RT$.

A problem of temperature was then to explain the constant quantity 273 *degrees Celsius*. That is, when $\tau = -273$ *degrees Celsius*, so that $T = 0$ *degrees absolute*, the ideal gas law becomes $PV = 0$. What is its meaning as it pertains to a real gas? The question was not addressed until the analysis of Ludwig Boltzmann (1844–1906) and his model of the gas in terms of a very large ensemble of atoms (or molecules).

It is interesting that Boltzmann was born in the same year that John Dalton (1766–1844) died. Dalton discovered the existence of distinct atomic weights. This followed from his discovery of the *law of partial pressures*.

What Dalton did was to study a mixture of gases, such as air (about 80% Nitrogen and 20% Oxygen). He used chemical means to evacuate from a box of such a mixture only one of its ingredients. He then saw that the pressure of the mixture of gases against its walls had a sudden change.

The pressure of the gas, by definition, is the perpendicular force per unit area that is exerted by the gas on the walls of its container. The sudden change of the pressure in Dalton's experiment then meant that the gas that was evacuated from the box, was *characterized by a discrete mass*, which no longer acted on the walls of the box. Thus Dalton learned that all of the atomic (or molecular) species have their own discrete atomic weights. It was then found that discrete atomic weights may be correlated with discrete atomic numbers: that of atomic Nitrogen N is 14, that of molecular Nitrogen N_2 is 28, the atomic number of Hydrogen H is 1 and that of molecular hydrogen H_2 is 2, and so on.

Amedio Avogadro (1776–1856) was a contemporary of Dalton. He discovered the 'mole' of a gas. He found that the weight of any atomic species, in grams, that is numerically equal to the atomic number of that atom or molecule, called a 'mole,' must contain the same number of constituent atoms (or molecules). It is called *Avagadro's number N*, which he found was the order of magnitude of 10^{24} atoms/mole. That is, one gram of hydrogen contains 10^{24} atoms of Hydrogen, 28 grams of molecular Nitrogen contains this same number of Nitrogen molecules, and so on.

Later in the 19th century, Boltzmann studied the question of the meaning of the temperature in the gas law. He saw that in fundamental terms, he would have to analyze the motion of 10^{24} atoms of matter — their mutual interactions and forces on the walls of the container, in terms of their Newtonian trajectories. Of course, for such a number this is an impossible task! So he invented the 'kinetic theory of gases,' whereby he was able to determine the *average values* of the physical properties of the atoms of the gas without the need to know their individual energies, momenta, etc.[19]

In this way, Boltzmann discovered that the average kinetic energy of a single constituent atom of a gas is $\langle(1/2)mv^2\rangle = (3/2)(R/N)T = (3/2)kT$, where k, the ratio of the universal gas constant to Avogadro's number, R/N, is called 'Boltzmann's constant.' Thus Boltzmann discovered that the gas temperature (in *degrees absolute*) is related to the average kinetic energy of the constituent atoms of the gas. This result then answered his question about the meaning of temperature, in terms of the dynamics of the atoms of the gas. Thus, at $T = 0$ *degrees absolute*, the average kinetic energy of the atoms of the gas is zero. That is, they are not (on the average) moving! As the temperature increases from this value, the atoms increase their speeds. But at $T = 0$ *degrees absolute*, according to Boltzmann's discovery, it is as though all of the constituent atoms of the gas are sitting in a pile at the bottom of the container. In this case, there is no force of the atoms on the walls of the container, so that the pressure of the gas, $P = 0$, and the volume of the atoms is infinitesimal compared with the volume of the container, so that $V = 0$.

Heat and the Conservation of Energy

During the 19th century, there was a great deal of discussion and research on the subject of heat. Some of the seminal work on this problem was done by James Joule (1818–1889), a student of John Dalton. He discovered the mechanical theory of heat — he was able to correlate the mechanical energy of a paddle turning in a bucket of water with the temperature increase of the water. He then concluded that the heat energy thereby generated in the water was converted from the mechanical energy of the paddle.

In this period of the history of science, the *law of energy conservation* was discovered. It is the assertion that among the different sorts of energy (potential, kinetic, mechanical, chemical, electrical, heat, ...) their total must be constant *in time*. An example is the consideration of a rock, initially sitting on the top of a cliff. At that time it has potential energy, by virtue of its height above the bottom of the cliff. When it is pushed from the edge of the cliff, in falling to the ground it continuously converts its initial potential energy into increasing kinetic energy (of motion), such that the sum of the remaining potential energy at any distance from the ground and its kinetic energy, at any time during the fall, is constant. Near the bottom of the fall, the rock loses all of its initial potential energy to a maximum kinetic energy and when it hits the ground this is converted to either heating the ground (heat energy) or the mechanical energy of breaking apart. Thus the total energy of the rock, that is constant in time, was initially all potential energy, and then a combination of potential and kinetic energy, is converted into the work required to break the rock apart or to heat the ground. The total energy of matter, *per se*, is then *defined to be the work that this matter is capable of doing. It is a quality of matter that is constant in time*. This is the *law of the conservation of energy*.

The Laws of Thermodynamics and Atomism

The *first law of thermodynamics* is a consequence of the conservation of energy. It is this: if ΔQ is the increase in the heat energy injected into a material system, ΔU is the corresponding increase in the internal energy of the system, and $P\Delta V$ is the change in the work exerted by the gas against the walls containing this system of matter, then $\Delta Q = \Delta U + P\Delta V$. [Of course, if the walls of the material system are infinitely rigid, then $\Delta V = 0$ and $\Delta Q = \Delta U$.]

The *second law of thermodynamics* is more subtle than the first. It concerns a quality of the gas called "entropy" — the measure of *disorder* of a complex system. The *second law of thermodynamics* asserts that if a complex system is initially in a state of non-equilibrium, then it will proceed *in time* toward the equilibrium state, as the entropy (disorder) of the system increases toward its maximum value, at equilibrium. The system will then remain in this state forever.

An example to illustrate this process was proposed by Josiah W. Gibbs, one of the discoverers of the subject of *statistical mechanics*. Suppose that a drop of dark blue ink is inserted into a beaker containing a clear liquid. At first, there would be maximum order in the sense that *one would know* where all of the ink molecules are located (i.e. in the confines of the drop of ink). But *in time*, the ink molecules would break away from the drop and they would diffuse into the clear liquid — until the entire drop of ink diffuses and the liquid becomes pale blue in color. The liquid would then stay in this pale blue state. In the latter state of equilibrium of the liquid with the ink in it, the disorder (entropy) is a maximum in the sense that *one would have minimal knowledge* of where any particular ink molecule would be located.

In this example, we see the idea of the 'arrow of time' — that time flows in only one direction, the direction of the increase of the entropy of the system, from non-equilibrium to equilibrium.[20] Here, 'time' is *defined* in terms of the *observation* of the evolution of the system from a more ordered state to less ordered state. [*This is not the same concept of 'time' that Newton introduced in his laws of motion — as the parametrization of the evolution of a* single body *along a trajectory, or the time that is involved in* the single trajectory *of a moving body in Einstein's relativity theory.*]

Boltzmann was able to formulate the two laws of thermodynamics in terms of the *statistical mechanics* of a many particle system of the atoms (or molecules) of matter. He invoked 'statistics' for the description of this matter. But he assumed all the while that *underlying* the statistical analysis, were the predetermined Newtonian trajectories of the atomic (or molecular) constituents of a macro-quantity of matter.

In the 20th century, the statistical aspect comes in a more fundamental way in the quantum theory. For the quantum theory is indeed a statistical theory of micromatter, in which it is asserted that a particular law of statistics (i.e. quantum mechanics) is itself a fundamental law of matter. Thus we see here a 'thread' in the use of statistics in physics, from a *subjective* meaning of matter (Boltzmann) to an *objective* meaning (according to Bohr, and his interpretation of quantum mechanics). In the latter view, the basic laws of matter are laws of statistics. This transition in the use of statistics in physics (a 'thread') will be developed later on in the course, in our discussion of the quantum theory.

In this regard, there is a question about the *fundamental nature* of the law of entropy (the second law of thermodynamics). For it is based on our human observations of the order of a system, not on the intrinsic order of the system itself. Similarly, in the 20th century, the 'observer' plays a seminal role in the laws of matter according to the view of quantum mechanics. Here, the trajectories of the atoms of matter are not predetermined. They are only specified after an 'observer' makes a measurement. It this sense, the quantum theory is non-deterministic. On the other hand, in the theory of relativity of the 20th century, the 'observer' plays no fundamental role and there is no objective 'arrow of time.' The laws of nature and the variables of matter are *predetermined*, according to relativity theory — which in this sense is *deterministic*.

A question that follows is this: Can it be meaningful to talk about the 'entropy of the universe'? I do not believe so because the universe is *by definition*, whatever else we may know about it, a *totally ordered* and a closed entity. It cannot be described in the context of disorder. It is only *our knowledge* of the universe that can be disordered. We should not confuse what it is that we know or do not know with what exists, independent of our knowledge of it (i.e. confusion between questions of epistemology and ontology)!

This is a monumental error, in my view, in modern day science — to confuse an objective entity, such as an electron, or the universe itself, with our knowledge of these entities! Some scientists say that it is a definition of science that it deals *only* with the empirical facts — our *description* of nature. But in my view, the basic goal of science is at a level underlying this — it is the *explanatory* level of understanding. Those who believe in the former as the goal of science are the *positivists*. Those who believe the latter are the *realists*.

Ludwig Boltzmann's and Ernst Mach's Philosophy of Science

Boltzmann's epistemological view was that of *naive realism*. This is the belief that the fundamental ways of the world that one responds to with one's senses are the same as those aspects of the world that one does not directly respond to. That is, because we see the world atomistically, as 'things,' he believed that any macromatter must be composed of smaller, unobservable things — the atoms of matter that we do not respond to directly.[21]

Thus Boltzmann concluded that the laws of nature are laws of the atoms of matter — their motions and mutual interactions (as unobservable constituents of macromatter).

Boltzmann's adversarial colleague, Ernst Mach, disagreed with him about the reality of atoms. He argued that the atom, an unobservable entity, is but a mathematical artifice whose only purpose is to facilitate a description of macromatter. His contention was based on his philosophical view of *positivism*. This view asserts that the only reality is our human reactions to the outside world.

Boltzmann's belief in the atomistic view of matter was strengthened by his success in representing a gas (or other macromatter) by a large number of composite atoms or molecules to underlie the laws of nature. As we discussed earlier, the absolute temperature T in the ideal gas law, $PV = RT$, was explained in terms of the *average* kinetic energy of a constituent atom (or molecule) of the gas containing about 10^{24} atoms, without the need to know the precise kinetic energy of any single one of them.

With this mindset, Boltzmann was able to formulate the two laws of thermodynamics in terms of such a collection of atoms with a statistical analysis of the system, i.e. using the methods of *statistical mechanics*. Along the way, he demonstrated the unidirectional character of time's arrow in a formulation of the entropy of a many-atom physical system, in accordance with the second law of thermodynamics.

Agreements between Boltzmann and Mach and 'Mach's Principle'

As I indicated earlier, Boltzmann and Mach disagreed on the correct epistemology for the philosophy of science — Boltzmann's naive realism and Mach's positivism. But they did agree that all of the laws of physics are in principle refutable. [*This is a view, propounded in our own day, by the philosopher Karl Popper — an avid follower of Boltzmann.*] Also, both Boltzmann and Mach were anti-dogmatic about the laws of nature; they did not believe that any of the principles that are alleged to underlie the ways of nature should be advocated dogmatically, as would be the attitude of a *religious truth* (as we discussed in Lecture I)! In particular, they both felt that one should always look for alternative interpretations of the descriptive aspect of physics that might equally explain phenomena.[22]

A good example was Mach's view against atomism in his interpretation of Newton's second law of motion, $F = ma$. According to Newton's atomistic view, his second law of motion comes from the empirical observation that if F_1 and F_2 are the magnitudes of two different forces *that act on the same body* (i.e. having the same mass m), that cause its different accelerations a_1 and a_2, then the empirical value of the ratio of forces is *linearly* equal to the ratio of caused accelerations, $F_1/F_2 = a_1/a_2$. Thus m, the intrinsic inertial mass of the atom of matter acted upon, is introduced by Newton as the constant of proportionality between F and a, i.e. $F = ma$. In the ratio of forces acting *on the same body*, the parameter m cancels.

But Mach argued that one may interpret the empirically correct formula, $F = ma$, by noting that for *two different bodies*, with masses, m_1 and m_2, *accelerating at the same rate* (e.g. in the case of different bodies that fall freely at the rate g, according to Galileo's law of gravity), the empirically correct result is: $F_1/F_2 = m_1/m_2$.

This result may then be expressed as the law: $m = KF$, where the constant of proportionality between the inertial mass and the *total external force* acting on it is K (say $K = m_2/F_2$, taken as a standard for all other measurements).

The meaning of Mach's equation (that is consistent with Newton's second law of motion), is that the inertial mass of any local matter is not intrinsic to that matter!

Rather, the inertial mass of any localized matter is *caused by* the total external force that acts on it. This is a non-atomistic view of matter. The interpretation of inertial mass of any quantity of matter, that defines it to be dependent on all other matter — *of a closed universe* — was named by Einstein, "The Mach Principle."[23] This principle signifies that the most distant stars of the universe must affect the mass of a local bit of matter. But it is *all of the matter* of the universe, not only the distant stars that contribute to the mass of any local matter! Indeed, the nearby matter also has (most predominantly) an effect on the mass of any local matter. This idea will be discussed toward the end of the course. It will be seen that this principle plays an important role in the theory of general relativity that was to emerge in 20th century physics.

The Continuous Field Concept[24]

The primary introduction of the continuous field concept in 19th century physics was that of Michael Faraday (1791–1867). His interpretation of the laws of electricity and magnetism were seminal in this discovery. Faraday started his study with a criticism of

Newton's interpretation of his laws of physics in terms of *action-at-a-distance.* He did not like this concept (nor, as we saw earlier, did Newton)! It is exemplified in Newton's law of universal gravitation, $F_g = - Gm_1m_2/R_{12}^2$. (The minus sign stands for an attractive gravitational force.) That is, if a body with mass m_1 is at the spatial distance R_{12} from the mass m_2, F_g is the force exerted by the latter mass on the former, dependent on the inverse square of their separation, and *independent of the time.* Thus, if a star, many light-years from Earth, should suddenly explode out of existence (*a 'light-year' is the distance traveled by light in a vacuum in one earth-year; light travels at about 10^{10} cm/sec, thus a light-year is about 10^{17} cm!*) we on earth would respond spontaneously, *without any time delay.*

Faraday had the following contrasting interpretation of Newton's law, in terms of *action by contact*: Call $P_1(R) = -Gm_1/R^2$ a continuous field of potential force that could be exerted by the body with mass m_1 on any other test body at the distance R from a center of force, except at the singular point $R = 0$. *The reason for the latter is that the source of the field of force acting on a test body cannot be where the 'test body' is located, by definition of this field of potential force. It is defined to be the effect of one body acting on a* different *body.* (*This interpretation eliminates the idea of 'self energy' that is prevalent in modern day particle physics.*) Faraday then said that when a 'test body' with mass m_2 should be placed at the location $R = R_{12}$ cm from the center of force of m_1, then the force acting on m_2 at that spatial point is: $F_g = m_2P_1(R = R_{12}) = -Gm_1m_2/R_{12}^2$, in agreement with Newton's formula.

Faraday's point was that it is not the discrete atom with its own electric charge and mass that is fundamental, with its *derivative feature* of a field of potential force extending throughout space. It is rather the continuous field of potential force $P(R)$ that is the fundamental way to represent this matter. The laws of matter are then not the laws of motion of discrete 'things,' as they are in Newtonian atomism. Rather the laws of matter are the laws of the continuous fields that are there to influence test bodies, wherever they are in space, relative to the influencing matter — it is *action by contact of a test body with a field of potential force, at the location of the test body* — replacing the notion of *action at a distance.*

Faraday applied this idea of field theory to the laws of electricity and magnetism. He saw that the configuration of an electrical field of potential force is different than that of a magnetic field of force. For example, the electric field associated with charged matter is a set of divergent lines, emanating from the place where the charge is located. The charge itself, in Faraday's view, is a manifestation of the field, rather than the other way around. The electric field is monopolar — based on either a positive electric charge or a negative electric charge, to be represented by the electric field. The magnetic field, on the other hand, is dipolar — it entails a north pole and a south pole, always together. The lines of the magnetic field then leave one of the poles and bend around to reach the other pole. (*If an ordinary bar magnet, with a north pole and a south pole at each end, is sliced in two, each of the pieces would still have a north pole and a south pole at each of its new ends — a north pole must always be accompanied by a south pole! Recent experimental work has been done to see if one might separate the poles of a magnet — to isolate a* 'magnetic monopole.' *There has been no confirmation* of *this speculation to this date.*) A great deal of experimental work on electricity and magnetism had already been done in Faraday's time. The electric field was studied extensively in the 18th century by the American statesman, Benjamin Franklin, and in France by Charles Coulomb, and in the 19th century by the French scientists, Ampere, Biot-Savart, the American scientist, Joseph Henry, ... and numerous other scientists worldwide.

A very impressive study was carried out in Denmark by Oersted. He found that an electrical current in a vertical wire induced a magnetic field of force in a plane that is perpendicular to the wire, as seen in the orientation of compass needles in that plane. Then, reversing the direction of the current, the compass needles all reversed their directions in the perpendicular plane. Since an electric current is an electric field of force in motion, this indicated that *a magnetic field of force is nothing more than an electric field of force in motion.* Faraday then recognized that, since motion is a subjective aspect of the observation, it must be just as true to say that *an electric field of force is not more than a magnetic field of force in motion*. Thus the objective field of force is not a purely electric field or a purely magnetic field. It is rather a *unified electromagnetic field of force*. It is only under special conditions of observation that this unified field appears to be a purely electric field, or under other special conditions of observation it appears to be a purely magnetic field. But the underlying field is a *unified* electromagnetic field of force. Thus, in the 19th century, Faraday introduced the idea of a unified field theory. It was not to be continued until the pioneering work of Einstein in the 20th century, when he tried to develop a unified field theory of electromagnetism and gravity. [*Unfortunately, Einstein did not succeed in this attempt in his day. However, there is strong reason to believe that both Einstein and Faraday (and Schrödinger, in Einstein's time) were right in this anticipation of a fully unified field theory to underlie the laws of nature.*]

During Faraday's later years, he met the young theoretical physicist, James Clerk Maxwell (1831–1879). Maxwell found a way to represent Faraday's findings in a mathematical expression of the laws of electromagnetism. But he saw that there was not a perfect symmetry in these laws that already expressed all of the known electromagnetic phenomena of his day. To symmetrize the equations, Maxwell added an extra term, called 'displacement current' that did not have empirical backing at that time. But this generalized form of *Maxwell's equations* did indeed give extra predictions about the physical world. It predicted all of the known optical phenomena — revealing distinct solutions that give the different colors of the visible spectrum, associated with different frequencies of electromagnetic radiation, as well as radiation solutions that yielded the predictions of (yet to be seen) infrared waves, radio waves, X-rays, gamma rays — all corresponding to different solutions of his field equations, associated with a spectrum of different frequencies. The Faraday–Maxwell field theory of electromagnetism also resolved the earlier dispute between the Newtonian view of optics, as a particle theory of light and that of Newton's contemporaries, Hooke and Huygens, who saw light as a wave phenomenon, proving the latter to be the true one.[25]

Maxwell's finding was a triumph for Faraday's continuous field concept, at least (thus far) in the fields of electromagnetism and optics, as well as the idea of a unified field theory. (Faraday tried, experimentally, but unsuccessfully, to unify the electromagnetic field theory with the laws of gravitation.) In fundamental terms, the continuous field concept was in competition with the atomistic views of the 19th century, of Dalton, Avogadro, Boltzmann, ... In the 20th century, it was suggested that this dichotomy might be resolved with the dualistic ideas of the quantum theory. But this was opposite to the fundamental basis of the theory of relativity, as we will discuss in later lectures.

In the next lecture we will discuss the anomalies that appeared on the scene at the end of the 19th century and the beginning of the 20th century, as dark clouds might appear in an otherwise clear blue sky. It was these anomalies that led to the two major revolutions in 20th century physics — the relativity and quantum theories of matter.

Lecture IV

EARLY ANOMALIES AND ELEMENTARY PARTICLES

The Perihelion Precession of Mercury's Orbit

It is important to realize that an "anomaly" in science is relative to the predictions of an ongoing theory (paradigm). When one theory in science is replaced with another, the anomalous behavior with respect to the old paradigm could become a natural behavior in the new paradigm. This is exemplified in the discovery in the 19th century that the planet Mercury does not move in a cyclic manner relative to the Sun. *It is not in a stationary orbit.* But this behavior is not anomalous in regard to Einstein's theory of general relativity, where stationary orbits are not generally predicted.

The observation of Mercury's orbit as not being cyclic was made by U. Leverrier (1811–1877) in 1859. This was in contradiction with Newton's theory of universal gravitation that predicts cyclic motion (when the Sun is the only source of the planet's motion). To resolve the problem, many astronomers cited the perturbing effect of the other sister planets of our solar system on Mercury's motion, that would cause such behavior, small as it would be compared with the Sun's influence. The locations of the other planets of our solar system, at all times, and their masses were known, so that the calculations could be made of the perturbing effect. It was then seen that it was not adequate to explain, in a quantitative way, Mercury's anomalous behavior. It was not until Einstein's theory of general relativity, in the 20th century, that predicted this non-stationary motion, that the anomaly became a natural behavior in the context of Einstein's theory of gravitation.

Stationary orbits appear in Newton's theory because of the separation of the space-dependent and the time-dependent parts of the solutions for motion. But in Einstein's theory, as we will discuss in a later lecture, space and time are not separable — there is only a single objective spacetime, where space (or time) measures in the expression of a law of nature in one reference frame become a mixture of space and time measures in the expression of the law in different reference frames. Thus stationary orbits of matter about other matter are not predicted, in agreement with the empirical evidence about Mercury's orbit. The noncyclic behavior of Mercury's orbit is equivalent to a *precession of the perihelion* of the elliptical orbit. The 'perihelion' refers to the point of closest approach to the Sun in the elliptical orbit, with the Sun's center at one of the focal points of the ellipse. This was Leverrier's observation — that it took more time in each of Mercury's years to return to the perihelion point of its elliptical orbit.

The lack of stationary orbits in the problem of cosmology also led to agreement with Hubble's observation of the expansion of the universe, wherein the galaxies of the universe are not in stationary orbits, but rather each is moving away from the others.

The Michelson–Morley Experiment

In the 19th century, James Clerk Maxwell (1831–1879) brilliantly structured the Maxwell field equations that underlie the explanation of all electromagnetic and optical phenomena known in his day.

While Faraday believed that the continuous field is the primary way to represent electrically charged matter, rather than being a derivative property of elements of matter, Maxwell believed that the electromagnetic field, such as the radiation that correctly describes light, is an excitation of an underlying material medium, called the 'ether.' Still, there was no direct evidence for the existence of Maxwell's postulated underlying ether. [*Newton also expressed belief in the existence of underlying ether to describe light propagation. Here, the constituent ether atoms were said to be displaced in the conduction of particles of light.*]

Abraham Michelson (1852–1931) and Edward Morley (1838–1923) designed an experiment to detect the ether, as the medium in which light propagates.[26] It utilized the Michelson interferometer, as its experimental tool. The idea is this: Two monochromatic light waves (i.e. each with the same frequency) are sent from a single source, propagating respectively parallel and perpendicular to the surface of the Earth. The first light wave moves parallel to the surface of the Earth to a mirror L cm away and reflects back to the source. Simultaneously, the second light wave, initially in phase with the first, is sent on a path perpendicular to the surface of the Earth, to a mirror L cm away and then reflects back to the location of the source. The two light beams then merge and are sent together to the interferometer, to detect their anticipated phase difference.

If the earth is moving through the ether at the speed $-v$ cm/sec (this is equivalent to the ether moving past the earth at v cm/sec) and light propagates at c cm/sec, then the time for the parallel propagation, to the mirror and back, is t (parallel) $= [L/(c + v) + L/(c - v)] = 2Lc/(c^2 - v^2)$ sec. For the perpendicular propagation, to and from the mirror, the total time is different; it is t (perpendicular) $= 2L/(c^2 + v^2)^{1/2}$ sec. [The denominator is the hypotenuse of a right triangle with sides c and v.]

Because of the different times of propagation of the two light beams, if they are initially in phase, it is expected that when they merge at the interferometer they will be out of phase, and the phase difference will relate to the speed of the ether, v, relative to the Earth.

But this expectation was not met! There was no phase difference detected, implying that $v = 0$. That is, it seemed that there was no ether there to conduct the light. This was an *anomaly*, in the context of the Maxwell theory of light propagation. In Michelson's thinking, it is as though a boat, moving at a speed c in a river that flows at the speed v would be expected to move at $(c - v)$ cm/sec upstream and at $(c + v)$ cm/sec downstream. The detection of no phase difference meant that there is no river flowing in the first place, i.e. that the boat is moving (in any direction) in a dry river bed. That is, it is as though there is no river to affect the boat's speed of motion. Their null result then showed that there is no need for an underlying ether to describe the propagation of light. This conclusion was in agreement with Faraday's concept of the fundamental nature of the field (of potentiality), not as a derivative property of a medium, but as the fundamental representation of electromagnetism and optics. It is also in accord with the theory of special relativity in correctly describing the motions of electrically charged bodies.

Blackbody Radiation and the Photon[27]

The experiment on blackbody radiation was one of the initiating studies that led to the quantum theory of the 20th century. Max Planck (1858–1947) correctly fit the curves for blackbody radiation, with the assumption that the energy of each frequency mode of a total radiation field is linearly proportional to its frequency, $E_f = hf$. h is a universal constant (with the dimension of 'mechanical action,' it is also the dimension of angular momentum). It is called "Planck's constant," with a value that is close to 10^{-27} erg-sec.

The experiment is the following: A cavity is placed in a constant heat bath, at a temperature T. The material of the cavity emits and absorbs radiation at an equal rate when the system of cavity and radiation are in thermodynamic equilibrium. One then inserts a filter in a small window in the wall of the cavity that will allow the passage of only one frequency component (frequency f and wavelength $\lambda = c/f$) of the enclosed radiation field. One then measures the intensity of this monochromatic radiation emerging through the filter. The measurement is then repeated for a whole spectrum of intensities corresponding to the different wavelengths in the radiation field contained in the cavity.

The data from this experiment is then plotted as a curve of intensity versus wavelength, yielding a bell-shaped curve. It is the 'blackbody radiation curve.' The experiment is then repeated for different temperature heat baths — all yielding the similar type of curve. The measurements are then repeated for cavities made of different materials. For the same temperatures, the different material cavities show the same blackbody radiation curves. The latter seems to indicate that the radiation field is an independent entity, not connected to the atomic constituency of the cavity, once it emits this radiation from its walls. It is a conclusion that is in contrast with Faraday's view — that the electromagnetic field of radiation corresponds to the dynamics of the charged matter of the cavity.

According to the classical radiation theory, one predicts that the left-hand side of the bell-shaped curve of intensity versus wavelength must go to infinity as the wavelength $\lambda \to 0$ (corresponding to the ultraviolet end of the spectrum; this is called the "ultraviolet catastrophe"). But the observation was that as $\lambda \to 0$, the intensity curve drops to zero. This 'anomalous' behavior was explained by Planck when he assumed that the energy in each frequency mode of the radiation field is linearly proportional to its frequency, $E_f = hf$, in his statistical analysis of the enclosed radiation field as a gas of distinguishable modes of radiation, at the temperature T. These modes were later identified as a gas of 'photons.' The photons were then taken to be bundles of 'quantized' electromagnetic radiation. Could these be the corpuscles of light that Newton referred to in his optical theory? The question then arises: Is the 'photon' an independent elementary particle, or is it rather a virtual field that is only a signal that is transferred between an emitter and an absorber to affect their electromagnetic coupling, but that it is not a *thing-in-itself*? [*Both Planck and Einstein came to the latter conclusion after their initial studies of the photon theory of light. Thus, in my view, they would have concluded that a 'photon' is not an elementary particle. The comments of Planck and Einstein on this question are in: "The Collected Papers of Albert Einstein Vol. 5: The Swiss Years Writings," 1902–1914, English Translation, Anna Beck; book review by M. Sachs*, Physics Today **48**, 65 (1995).] In the analysis of blackbody radiation by Planck, the photon is assumed to be a localized particle. But in other sorts of experimentation, the photon appears to be a wave. In these early days of the 'old quantum theory,' Einstein proposed that the photon is whatever one sees it to be in experimentation — a wave and a particle are both correct in this view, so long as one does the experiments to see it as 'wave' or 'particle' at different at times. It is a view that is consistent with the philosophical approach of *logical positivism*. This view came to be known as 'wave–particle dualism.' It is an idea that Einstein abandoned soon after this period, but it was applied in the 20th century (in quantum mechanics) to the nature of the photons and to the material particles, such as the electron.

The Electron

Near the beginning of the 20th century, J. J. Thomson discovered the electron — the lightest known elementary particle.[28] His was an experiment that entails the use of the 'cathode ray tube.' A voltage is generated at a negatively charged plate, at one end of the tube. It then emits a stream of electrons, toward a positive plate further on in the tube. The electron beam is sent through a series of controlled electric and magnetic fields, until it lands at a spot (determined by the electric and magnetic fields) on a phosphorescent screen at the other end of the tube.

From this data, J. J. Thomson was able to determine the ratio of charge to mass, e/m, of the electron. In a later experiment, Robert Millikan (1868–1953) measured the charge of the electron.[29] What he did was the following: He (singly) ionized oil drops within an electric field, E, between charged plates with an imposed voltage V between them. A single oil drop (with electric charge e) then falls in the gravitational field of the Earth, by virtue of its weight. By adjusting the electric field to oppose the gravitational force, to leave the oil drop at rest, Millikan was able to determine the electric charge, e, of the singly ionized oil drop, from the equality of the electric and gravitational force acting on the oil drop, i.e. the balance of gravitational and electric force on the oil drop: $eE = Mg$. M is the mass of the oil drop and E is the electric field, derived from the imposed voltage $V = Ed$, where d is the distance between the plates. With this knowledge of the electric charge e of an electron, and J. J. Thomson's measurement of e/m, it was then possible to determine the mass m of the electron. It turned out to be the order of 1/1840 of the mass of a proton — the nucleus of hydrogen. [*It is still a question in elementary particle physics,* why *the ratio of the mass of a proton to that of an electron is near 1840.*]

The Quantization of Electrical Charge

It is an experimental fact that the electrical coupling between any of the elementary particles is proportional to ne^2, where n is an integer: 1, 2, 3, ... and e^2 is the (square of) the electrical charge of the electron (as discovered by Millikan). This led to the implication that the elementary particles themselves must have an electric charge in integral units: 0, e, $2e$, ... such as the neutron, proton, helium nucleus, etc. In contradiction with this conclusion is a present day theory in elementary particle physics, called "the Standard Model," based on the makeup of nucleons (protons and neutrons) in terms of composites of fractionally charged particles, called "quarks." These have electric charges $\pm e/3$ and $\pm 2e/3$.

The conclusions of the fractionally charged quark model are based on indirect experimental evidence. There has been no direct experimental evidence, to this date, for the fractionally charged particles of matter, analogous to the Millikan oil drop experiment.

A brilliant experiment was carried out recently to look for fractionally charged particles of matter. It was a more sophisticated version of the Millikan oil drop experiment. [I. T. Lee, S. Fan, V. Halyo, E. R. Lee, P. C. Kim, M. L. Perl, H. Rogers, D. Loomba, K. S. Lackner and G. Shaw, "Large Bulk Matter Search for Fractional Charge Particles", Physical Review D66, 012002 (2002), Stanford Linear Accelerator Center, Pub. 9191, June, 2002.] According to the Abstract of this paper, "no evidence for such particles was found, giving a 95% confidence level upper limit of 1.7×10^{-22} particles per nucleon."

There is a group theoretical prediction in the holistic field theory of elementary matter in general relativity that indeed the integral coupling of elementary matter, e^2, $2e^2$, $3e^2$, ... is rigorously true

(Ref. 68, Sec. 7.8) — in agreement with the experimental findings of I. T. Lee *et al.*

The Photoelectric Effect[30]

There were several experiments near the beginning of the 20th century, in addition to blackbody radiation, that indicated a critical appearance of Planck's constant h. One of these was the photoelectric effect. The experiment is the following: A photosensitive piece of metal is placed (in series) in an electric circuit with a resistor, a voltmeter across the resistor, a battery and an ammeter. The ammeter reads the electric current I that flows. Monochromatic radiation (a single frequency) shines on the photosensitive metal. By increasing the frequency of the impinging light, a threshold value is reached when extra current in the circuit is generated. By increasing the frequency further, the current correspondingly increases. The *change of energy* of the moving electrons that constitute the current with increasing frequency is as follows: $\Delta E = e\Delta V = eR\Delta I = h\Delta f$, where ΔV is the corresponding change of voltage (in volts), across the resistor, due to the change of frequency of the impinging light, Δf cycles/sec, R is the resistance of the resistor (in ohms) and ΔI is the increase in current in the circuit (in amperes). That is, the change in current with changing frequency of the impinging light, is $\Delta I = (h/eR)\Delta f$ amperes. This is the "photoelectric effect." [*It was first analyzed by Einstein, who received the Nobel Prize for this research.*] Thus we see here the appearance of the same constant, Planck's constant h, that appears in Planck's formula for blackbody radiation.

The Compton Effect[31]

Planck's constant h also appears to explain the experiment on the scattering of monochromatic radiation (a photon) from an electrically charged body (an electron). The photon scatters from the charged body, at some angle, with less energy, and thus a smaller frequency, while the electron absorbs this energy and moves away in a definite direction. With the conservation of momentum and energy, calculation shows that the observed angle of the scattering of the charged body (an electron) is dependent on Planck's constant h, in agreement with the empirical observations.

Radioactivity

Near the beginning of the 20th century, it was discovered (accidentally!) by Antoine Becquerel (1852–1908) that there are some mineral elements (such as Radium) that emit energy on their own.[32] These are intrinsically unstable elements, emitting three types of radiant energy: alpha, beta and gamma rays. It was found that the alpha rays are the nuclei of Helium — they are positively, doubly charged particles with (approximately) four times the mass of a proton. Later in the 20th century, the Helium nucleus was discovered to be two protons and two neutrons, bound by nuclear forces, where the neutron is a charge-neutral particle whose mass is close to that of the proton. [*The elementary particle, 'neutron' was discovered by James Chadwick (1891–1974) in 1932*[33].] The beta rays are electrons (discovered later to be emitted in the beta decay of unstable nuclei). The source of this decay is called the 'weak interaction.' The gamma rays are electromagnetic radiation (photons). Such elements that emit these types of radiation are 'radioactive.'

Together with Marie and Pierre Curie (1867–1934) and (1859–1906), Becquerel found that many of these unstable, heavy elements may transmute into other of the heavy elements of the periodic table. This was the beginning of the observations of the elements of nuclear matter that are unstable. It was found that in their natural bulk state, many of these elements have isotopes (elements with the same number of protons but different numbers of neutrons) that are radioactive. For example, the stable element Uranium, U^{238}, that is made up of 92 protons and 146 neutrons, occurs in its natural bulk state containing a small fraction of its isotope U^{235} (92 protons and 143 neutrons) that is highly radioactive. To utilize the latter radioactive isotope, as an energy source, it is necessary to chemically separate it from the bulk U^{238}. A great deal of the work of Marie and Pierre Curie was to learn how to carry out this separation.

Rutherford and Atomic Structure[34]

Ernst Rutherford (1871–1937) discovered that the structure of the atomic elements is not as it was previously thought — a solid, positively charged material with negative electrons circulating throughout, to give it a net zero electric charge. His experiment was to scatter alpha particles from the atoms of a gold foil. He expected to see, on an absorbing screen, the shadow of the (supposedly) solid gold atom. Instead, he found that the alpha particles went through the gold atoms most of the time. But a small fraction of them bounced back from a tiny part of the gold atom, in its center. He saw that the radius of the reflecting center of the gold atom was about 10^5 times smaller than the radius of the atom itself (the latter could be determined in X-ray analyses). Thus, Rutherford discovered that most of the volume of an atom is empty space! Most of the mass of the atom is in this nucleus that is in its tiny central region, and it is positively charged. On the outside of the atom are the negatively-charged electrons that balance the positive charge of its nucleus, to make up a charge-neutral atom. [*If a human body could be kept intact while squeezing out all of its empty space, it would become 10^{15} of its normal size. Even an electron microscope would not be able to see it*!]

Bohr's Atomic Model

After Rutherford's discovery of the structure of the atom, Niels Bohr (1885–1962) arrived in Manchester, UK, to work with him as a research assistant. To explain Rutherford's result, Bohr proposed a planetary model of the atom, similar to the solar system of planets orbiting the Sun.[35] But instead of the attractive gravitational $1/r^2$ force that binds the planets to the Sun, it is the electric $1/r^2$ Coulomb force that binds the electrons to the nucleus of the atom.

A trouble with Bohr's atomic model is that, according to classical, electromagnetic theory, an accelerating electric charge, as is the case of the orbiting electric charge, must radiate energy away. In this case, the electron could not stay in a fixed orbit; it would spiral into the nucleus of the atom. Thus, such an atom could not be stable. Bohr resolved the problem by postulating that the angular momentum of the orbiting electron must be 'quantized,' in units of $h/2\pi$. (The hint for this quantization may come from the fact that the units of Planck's constant h are the same as those of angular momentum.) Thus, the classical orbital angular momentum, mvr, of the electron was taken by Bohr to be $nh/2\pi$, where $n = 0, 1, 2, \ldots$, and where v is the speed of the orbiting electron, and r is its radial distance from the nucleus.

Thus, with Bohr's model of the atom, the electron in a fixed orbit is in a particular *energy level* of the atom, E_n, until it would (acausally) 'jump' to a different energy level. When this 'jump' happens, to a lower energy level, the energy lost by the electron (and therefore the host atom), when it descends to the lower energy level E_m, is taken up in the creation of a photon with frequency f_{nm}, such that $E_n - E_m = hf_{nm}$. According to Bohr this is the explanation for the emission of monochromatic light.

If the atomic electron, in the elevated state E_n drops to the minimum energy state of the atom — its ground state E_0, in a number of steps, then

$$E_n - E_0 = (E_n - E_1) + (E_1 - E_2) + \cdots + (E_{m+1} - E_m) + (E_m - E_0) = \Delta E_{n0} = hf_{n0}.$$

Since $\Delta E_{n0}/h = f_{n0}$, it follows that by dividing the above equation by h,

$$f_{n0} = f_{n1} + f_{12} + \cdots + f_{m0}.$$

That is to say, the prediction of Bohr's model is that the frequency of any line in an atomic spectrum is the sum of frequencies of other lines in the same spectrum. This is the *Ritz Combination Principle*. It was discovered empirically in 1908, before Bohr's 1913 proposed atomic model of the atom. This theory of the atom is based on "the old quantum theory."

In the next lecture, we will discuss some of the further difficulties with Bohr's model of the atom, in the context of the "old quantum theory," such as the "quantum jump," and later results that led to the "new quantum mechanics."

Lecture V

FROM THE OLD QUANTUM THEORY TO QUANTUM MECHANICS

Bohr's Atom and Energy Levels

In the last lecture we discussed Bohr's planetary model of the atom. This followed from Rutherford's discovery that the atoms of matter are mostly empty space, with a positively charged nucleus, carrying most of the mass of the atom in a tiny fraction of the volume of the atom, at its center, and the negatively charged electrons circulating on the periphery of the atom. The key point that resolved the classical prediction that the atom would be unstable because the accelerating electrons in orbit would lose energy, according to Maxwell's theory, was Bohr's introduction of a new law — 'quantized' angular momentum for the orbital atomic electrons. This, in turn, led to the description of atoms in terms of a discrete energy level structure.[35]

The data that was used to verify the Bohr model is based on spectroscopy. What one does here is to first excite a gas, say Hydrogen, enclosed in a tube, by applying a voltage to it, and then observing its glow. (Or one might look at a star — a 'furnace' that is already made up of excited atoms and molecules.) This light from the excited gas (or star) is then passed through a prism that separates out its different frequency components. These appear as the 'lines' of a spectrum, characteristic of this particular atomic type. [*A way to determine the constituency of a given star is to examine its spectra. Each atomic or molecular constituent has a characteristic line spectrum. These spectra may then be used to identify the atomic makeup of the star.*]

According to Bohr, the generation of light of specific frequencies comes from the electron in the higher energy level E_n dropping to a lower energy level E_m. This leads to the simultaneous creation of a photon with frequency f_{nm}, with the photon energy $hf_{nm} = E_n - E_m$.

As we discussed in the previous lecture, if an electron drops from the energy level E_n to a lower energy level E_0 in several discrete jumps, there is the implication of the empirically confirmed *Ritz combination principle*: $f_{n0} = f_{n1} + f_{12} + \cdots + f_{m0}$. That is, the frequency of any particular line in the spectrum is a sum of frequencies of some of the other lines in the same spectrum.

But there still remain problems with this model and the concept of the 'quantum jump.' The first problem is this: Why does the electron stay in the excited energy level for an unspecified period (except for an average amount of time) before it descends to a lower energy level? What is the physical mechanism that makes it wait? In contrast, in classical physics if a round stone sits on the side of a hill, propped by another stone, and the second stone is then removed, the round stone will roll down the hill, *immediately*. It is because of the action of the force of gravity that is there all the time. In comparison, the electron in the excited state of an atom does not drop to the lower energy level because of the electric force of the nucleus on it. There is no answer to this question in the context of classical physics.

A second question concerns the law of the conservation of energy. When an electron in an excited energy level E_n drops to a lower energy level E_m, to conserve energy a photon is created with the same energy as the loss of energy of the electron (and its host atom). The problem is this: After the electron drops from the energy level E_n, but *before it arrives* at the energy level E_m, it will have lost energy, but there is yet no photon to take up this loss! Thus, with this view, the law of conservation of energy does not hold true at these interim times of transition between states. It is only true when the electron is in one definite state or another.

A third problem is this: What is the physical mechanism that *causes* the creation of a photon, when the electron arrives at the lower energy level?

In Bohr's theory, one merely says that these things happen, without the need to specify physical causes. It is similar to the attitude of Aristotle when he said that a body falls from an *up* position to a *down* position, because *down* is the natural place for the body! He did not specify any cause for the fall, as Galileo and Newton did, 2000 years later.

These difficulties with the Bohr atom then led to the need for advancement to a higher stage of understanding of atomic matter. It eventually led to the formulation of "quantum mechanics." A giant intermediate step toward quantum mechanics was the empirical discovery of the wave nature of matter.

Electron Diffraction and Wave–Particle Dualism

In 1923 Louis de Broglie (1892–1987) postulated that material particles, such as electrons, have a wave nature. The particle-like variable, its momentum p, relates to the wave-like variable, its wavelength λ, according to the reciprocal relation $p = h/\lambda$. [This is a generalization of the reciprocal relation for the photon. While the photon has no inertial mass, it does have momentum $p = E_\gamma/c = hf/c = h/\lambda$.]

Four years after de Broglie's postulation, in 1927, C. J. Davisson and L. H. Germer (1881–1958) and (1896–1971), in the US, and simultaneously in the UK, G. P. Thomson (1892–1975) (the son of the discoverer of the *discrete* electron, J. J. Thomson), discovered that the scattering of an electron from a crystal lattice yields a diffraction pattern.[36] That is, they saw on an absorbing screen of the scattered electrons, instead of revealing the geometrical shadow of the crystal lattice, as discrete particles would, it revealed a constructive and destructive interference pattern. The latter are the respective locations on the absorbing screen where electrons bunch together and locations where no electrons land. That is, in this experiment, the material particles, electrons, *revealed themselves to be continuous waves rather than discrete particles.*

This result was one of the most important experimental findings of 20th century physics. It showed that the elements of matter are continuously distributed fields rather than the atomistic model of matter as a collection of point particles. How did the physicists of the day explain this experimental finding *while still insisting that electrons are fundamentally discrete particles*? What they did was to evoke the *principle of wave-particle dualism,*[37] originally suggested (and later revoked) by Einstein in his explanation for the quanta of light — the photons. Bohr and his followers (the 'Copenhagen School') said this: Under the experimental conditions designed to see the electrons as discrete particles, they are discrete particles at that time. But under the experimental conditions to see the electrons as continuous waves, they are continuous waves at that time!

In other words, the 'true' nature of the electron (and all other elementary particles of matter) is dependent on how a macro-observer chooses to see it — even though the discrete particle and continuous wave models are logically dichotomous. [This view, taken by the Copenhagen School, is compatible with the philosophical stand of *logical positivism.*]

To say that an elementary particle, such as the electron, is whatever it is, wave or particle, independent of any observations, is the philosophical view of *realism* — a view that was taken by Einstein and Schrödinger, and very few others.

Schrödinger's Wave Mechanics

The next step in our understanding of microscopic matter was to find the physical law that determines the wave nature of matter. Erwin Schrödinger (1887–1961) discovered the equation whose solutions are the discovered matter waves. [*The "Schrödinger wave equation" was discovered by generalizing the Hamilton–Jacobi equation of classical physics. This was a transformation from the nonlinear equation whose solutions are the 'action function' of classical systems, to a linear wave equation. It is analogous to the transformation from the description of '(geometrical) ray optics' to 'physical optics,' in terms of propagating waves.*]

The linearity of the Schrödinger wave equation implies that the sum of any number of solutions is another possible solution of the same equation. This feature was then classified as a principle of microscopic matter — the *principle of linear superposition*. Linearity is also a feature of a probability theory. Thus, there was a *possible* connection between Schrödinger's wave mechanics and a probability calculus. In this way, Schrödinger's wave mechanics was taken by the 'Copenhagen School' to be an expression of a probability calculus, 'quantum mechanics,' to underlie the nature of elementary matter.

This was shown by Max Born (1882–1970). He demonstrated that one could express the set of solutions of wave mechanics as the elements of a 'Hilbert space,' that is amenable to a probability interpretation. Here, one has an infinite sum of absolute squares of the solutions of the wave equation that is equal to unity — just as the sum of all probabilities, *by definition*, must add up to unity. Geometrically, this is analogous to the three-dimensional (rather than infinite-dimensional) invariance of the radius r of a sphere under rotations, according to: $x^2 + y^2 + z^2 = r^2$. The coordinates, x, y and z each change under rotations about an axis through the center of the sphere, but the radius r stays invariant. Quantum mechanics is, however, not an ordinary probability theory. In the classical probability theory, variables relate to the probabilities of particular states; in quantum mechanics, it also includes the probabilities of transitions between states. With this interpretation of the wave function, it then became possible to determine the average values of all of the physical properties of microscopic matter, when it is in any of its possible states, as well as determining the probabilities of transitions between the different states of elementary matter.

Schrödinger's Interpretation of Wave Mechanics

Schrödinger himself did not accept Born's probability interpretation of his wave mechanics. He saw the matter waves as real waves that are the normal modes of vibration of a collection of bodies, but not tied to any particular one of these bodies. It is analogous to a large ensemble of material blocks connected with springs. By initiating the motion of one of the blocks, one way or another, the entire ensemble vibrates with characteristic frequencies. This was Schrödinger's interpretation of the matter waves. Born, on the other hand, saw the matter wave as tied to a single particle of matter.

What Schrödinger had in mind was that the wave function (a solution of his wave equation) — a complex function to represent a propagating wave — relates to the real number charge density ρ, on the right-hand side of Maxwell's equations (and the corresponding current density). This gives rise to the electric and magnetic field variables on the left, by identifying the charge density $\rho = e\psi^*\psi = e|\psi|^2$ Coulombs/cm^3, where e is the charge of the electron (in Coulombs) and $|\psi|^2$ is the 'probability density.' [*Thus the dimension of the wave function* ψ *— called 'probability amplitude' — is* 1/cm$^{3/2}$.] It is only in the micromatter domain where this product of the wave function and its complex conjugate unfolds, to reveal the (complex number) wave function ψ by itself — its manifestations in terms of constructive and destructive interference.

Schrödinger's wave mechanics and its expression in terms of a probability calculus — corresponding to a 'Hilbert space' — then gave the name "quantum mechanics" to this new view of elementary matter.

A problem that arises with Schrödinger's view is this: How does one atom of matter — an emitter — transfer its vibrational energy to another atom — an absorber — so suddenly and sharply? Schrödinger saw this in terms of a resonance condition. It is analogous to the following model. A block, B1, is coupled vertically to the ceiling with a stiff spring. It is coupled horizontally to another block, B2, with a weak spring. B2 is also coupled to the ceiling, vertically, with a stiff spring. At first, B2 is at rest when one pulls B1 from the ceiling to start its rapid oscillation. Eventually, B1 transfers its oscillations to B2 via the weak spring coupling to it. In time, B1 slows down, and then stops its oscillations while B2 absorbs these oscillations. The process then reverses and B1 starts to speed up its oscillations and B2 slows down, until B1 is again at the maximum amplitude of its oscillations and B2 stops, *and so on ad infinitum.* This is called a 'resonance condition.'

Such a resonance condition is expressed as follows: The loss of frequency of an emitter is equal to the gain in frequency of an absorber: $f_e - f_{e'} = f_{a'} - f_{a'}$ $f_{e'}$ is the decreased frequency of the emitter and $f_{a'}$ is the increased frequency of the absorber.

Multiplying this equation by Planck's constant h, with $E = hf$, we have: $E_e + E_a = E_{e'} + E_{a'}$. This is a statement of the conservation of energy from the time before the oscillation was transferred to the time after it was transferred. If there would be many absorbers, then the loss of frequency of the emitter to all of the absorbers is $(f_e - f_{e'}) = (f_{a'} - f_{a(1)}) + (f_{a(1)} - f_{a(2)}) + \cdots + (f_{a(n)} - f_a)$. That is, there is a 'total frequency' f_t, such that we have the conservation law: $f_t = f_e + f_a + f_{a(1)} + \cdots = f_{e'} + f_{a'} + f_{a(1)'} + \cdots$. This describes the resonance condition that prevails in the description of the transfer of oscillations from an emitter to an absorber, in a causal fashion, according to Schrödinger's model. The rapidity and sharpness of the transfer of the oscillations depends on the physical properties of the coupled atoms of matter, analogous to the properties of the blocks and springs in Schrödinger's model.

The Young Double Slit Experiment

If the electron is truly a continuous wave, then if it approaches a screen with two open slits in it, the transmitted electron wave to a second screen beyond the first one should reveal a diffraction pattern. This was originally analyzed by Thomas Young (1773–1829), in his consideration of the wave propagation for light. In this way, Young proved the wave nature of light (in disagreement with the earlier corpuscular model of light according to Newton and in agreement with the wave model of Newton's contemporaries, Huygens and Hooke). In that period it was found that light is a transverse wave, with the vibrations perpendicular to the direction of propagation, while Huygens believed that light is a longitudinal wave, such as sound waves. In the case of electron diffraction, the Young double slit experiment is a *thought experiment* for material particles (such as electrons) that *corresponds to* the real observation of electron diffraction from a crystal lattice.

Suppose, instead of a purely wave model for the electron, we accept the probability interpretation applied to individual discrete particles, with the wave function solution from quantum mechanics. Thus, we consider a screen S_1 with two slits in it, s_1 and s_2. There is a second screen S_2 beyond the first screen. Sending the electron beam to S_1 with slit s_2 closed, the probability density for the electron going through slit s_1 is $|\psi_1|^2$. This shows up as an (almost) sharp image of s_1 on the screen S_2. (It is 'almost' sharp because there is some diffraction of the electron wave as it bends *slightly* around the corner of slit s_1, because it is a continuous wave, not a discrete particle.) Similarly, if the slit s_1 is closed, the probability density for the electron wave to go through slit s_2 is $|\psi_2|^2$, revealing the 'almost' sharp image of slit s_2 on the second screen S_2. But if both slits are open, the wave function for the electron penetrating screen S_1 is the superposition of states, $(\psi_1 + \psi_2)$, so that the probability density for the electron wave reaching screen S_2 is $|(\psi_1 + \psi_2)|^2 = |\psi_1|^2 + |\psi_2|^2 + (\psi_1^* \psi_2 + \psi_1 \psi_2^*)$.

The first two terms above are the partial probability densities that the electron will pass through slit s_1 or slit s_2. The third term (the cross product) is the 'interference' part of the scattering. It shows up on screen S_2 as a *diffraction pattern*. The maximum of this pattern appears opposite the solid part of S_1, between the two open slits. On its right and left are a series of interference minima (destructive) and maxima (constructive), with decreasing amplitudes.

This seems peculiar if we are truly describing real electrons as discrete particles of matter. But if, indeed, electrons are pure waves (as Schrödinger interpreted), and not particles at all, then the diffraction pattern seen on the absorbing screen S_2 is exactly what is expected, as originally observed by Young in his studies of light propagation.

It should be mentioned that when the electron is claimed to be seen as a particle, as in J. J. Thomson's cathode ray experiment, it is not really seen as a truly discrete (point) particle — this is impossible. What is seen is a small but *fuzzy* spot on the phosphorescent screen. If one should look into this spot with a high enough resolution, a diffraction pattern would be seen inside of it. Thus, this is, empirically, a 'bunched' wave and not a discrete particle in the first place. [In the quantum theory, as we will discuss in the next lecture, this observation is attributed to the *Heisenberg uncertainty principle.*]

Einstein's Objection to Born's Interpretation of Linear Superposition

Einstein proposed the following thought experiment[38] to refute Born's interpretation of wave mechanics: Consider a hemispherical film, with a hole at its geometric center. An electron approaches the hole. Without looking for the electron in the hemispherical, region, an 'electron field' must, according to Born, include all possible states of motion (i.e. in all directions), described with the waves, ψ_1, ψ_2, ...ψ_P, ... However, as soon as one looks at the hemispherical film, one sees that the electron had landed at a particular spot, P, described with a single wave function, ψ_P. That is, by looking at the film, all of the other wave functions must have collapsed to this one, ψ_P, *spontaneously*. Einstein then said that this claim would contradict the prediction of the theory of special relativity (to be discussed in a later lecture), that there can be no spontaneous action at a distance. It must take a finite time for the waves at all other places on the hemispherical film to collapse to the single point at P.

Born then replied that these waves we discuss, landing at all possible places on the hemispherical film, refer only to *our knowledge* of where the electron is located. When we see that it landed at the specific place, P, our knowledge becomes certain. This is the meaning of the collapse of the wave packet. It is a *subjective meaning*, not at all concerned with the electron as a thing in itself. *This idea of an intrinsic subjectivism in the definition of the elementary particle of matter was a revolutionary aspect of the Copenhagen School's interpretation of quantum mechanics as a theory of matter that is dependent on its measurement by a macro-observer.*

Schrödinger's Cat Paradox[39]

Finally, Schrödinger objected to Born's claim that the superposition of states only refers to micromatter and not to macromatter. He proposed the following thought experiment to refute this claim: Consider a box and a vertical wall in its center, with a hole in it. On one side of the wall there is a piece of radioactive matter emitting alpha particles. Opposite the hole in the wall, on its other side of the box, there is a Geiger counter, connected to a solenoid and, in turn, to a hammer that is pointed at a bottle of poison. Sitting in this side of the box there is a live cat.

The alpha particles are being randomly emitted. According to Born's view of quantum mechanics, any given alpha particle is represented by a superposition of states of going through the hole in the wall and not going through this hole. If the alpha particle goes through the hole in the wall, it triggers the Geiger counter to excite the solenoid to trip the hammer to break the bottle of poison. This would mean that the cat would die. If the alpha particle does not go through the hole in the wall, the cat would live. Thus, the superposition of the states of the (microscopic) alpha particle in going through the hole in the wall and not going through this hole would correlate with the states of life and death of the (macroscopic) cat. That is, if one chooses not to look into the side of the box where the cat is, it would be in a superposition of states of life and death — a logical paradox! But as soon as one would look into the cat's side of the box, one would see that it is either dead or alive! Schrödinger then saw that this conclusion was not logically defensible.

Born's answer to Schrödinger's refutation was similar to his answer to Einstein's thought experiment. It was to say that the wave function relates *subjectively* to *our knowledge* of the state of life or death of the cat, not to its objective state, independent of our knowledge. The superposition of states of life and death of the cat simply means that we did not know, before looking into the box, whether the cat was dead or alive!

Still, Schrödinger's cat paradox thought experiment did show, contrary to Born's claim, that there must be a correlation between the superposition of states of a microbody (the alpha particle in this case) and a macrobody (the cat). He then claimed to have refuted Born's interpretation of the linear superposition principle of quantum mechanics.

In the next lecture we will discuss Heisenberg's formulation of quantum mechanics and the Copenhagen School's interpretation in terms of the fundamental role of the measurement in the definition of the elements of matter.

Lecture VI

QUANTUM MECHANICS: HEISENBERG'S MATRIX MECHANICS AND THE COPENHAGEN SCHOOL

Heisenberg's Philosophy

Werner Heisenberg (1901–1976) was an early and significant collaborator of Niels Bohr. He was an even stauncher defender of the epistemological view of *positivism* than was Bohr. In one of his introductory articles on quantum mechanics, Heisenberg said: "The present paper seeks to establish a basis of theoretical quantum mechanics founded exclusively upon relationships between quantities which in principle are observable."[40] Ernst Mach would have been overjoyed to read this statement. That is, with Mach, Heisenberg believed that there is nothing meaningful in science that 'does not meet the eye!'

After Bohr's discovery of atomic structure in terms of discrete energy levels, Heisenberg voiced criticism of the model because 'we don't directly see the atomic electrons in their orbits.' He then started his program to formulate the properties of atoms strictly in terms of the observables — Bohr's energy levels and the transitions between them. This led him to the formulation of 'matrix mechanics' — a set of properties of the atoms of matter that was expressed in terms of matrices of numbers and the 'matrix algebra.' This was very successful in predicting the empirical atomic phenomena, as successful as was Schrödinger's wave mechanics.[41] A short time after Heisenberg's discovery, Schrödinger (and independently, Cornelius Lanczos (1893–1974)) showed[42] that wave mechanics and matrix mechanics are mathematically equivalent — these two formulations of quantum mechanics may be transformed into one another. [*The transformation of Schrödinger's wave mechanics to Heisenberg's matrix mechanics was a matter of converting a 'differential equation' into an 'integral equation.'*] Heisenberg was not happy with this! He thought of his formulation in terms of pure numbers to express the observables as a unique formulation of quantum mechanics. On the other hand, Schrödinger's formulation was in terms of an abstract field (the 'wave function' — that Heisenberg considered to be 'metaphysical!') that only after analysis yielded the predictions of atomic phenomena.

Matrix Mechanics

The first quantum mechanical concept that we must understand at the outset, in either the formulation of wave mechanics or matrix mechanics, is that of the 'state.' With Bohr's view, there is a distinct set of states of an atomic system, labeled by a discrete set of integers: 0, 1, 2, ..., n, ... In Schrödinger's formulation, these discrete states are the distinct set of wave functions: $\psi_1, \psi_2, \ldots, \psi_n, \ldots$

> [*In the following paragraphs there will be an unavoidable amount of mathematical expression. For those who are unacquainted with this terminology, these paragraphs may be skipped without loss of the main meaning expressed in this lecture.*]

Heisenberg's idea is this: If A represents the measurement of a particular property of elementary matter, then he introduced the number A_{nm}. When $n = m$, then A_{nn} is a measurement of the property A of the system, when it is in the state n. (These are the 'diagonal' elements of a matrix.) The off-diagonal elements of this

matrix of numbers, $A_{n \neq m}$ represent a transition between the states n and m, induced by the physical property A. Thus the full set of 'observables' of A, distributed among the states of the system of states, is the collection of numbers, A: $(A_{11}, A_{12}, A_{13}, \ldots, A_{1n}, A_{21}, A_{22}, A_{23}, \ldots, A_{nm}, \ldots, A_{nn})$.

The measurement of another property of the system, B, is similarly represented by the collection of observable numbers, B: $(B_{11}, B_{12}, \ldots, B_{nn})$. The succession of the measurements of A and B is then given by the matrix of numbers: $(AB)_{nm} = C_{nm} = A_{n1}B_{1m} + A_{n2}B_{2m} + \cdots = \Sigma_k A_{nk}B_{km}$, where the latter is a sum taken over all 'intermediate states' k. In short, for the description of a succession of measurements, A and B, we have the numerical representation $C = AB$, according to the product rule for matrices, shown above.

It is important that $AB \neq BA$. The product rule for the matrices of numbers is just the rule that was discovered by A. Cayley (1821–1895) when he invented the noncommutative *algebra of matrices*. Thus, Heisenberg's theory of atomic systems is called "matrix mechanics." [*It was Max Born who taught Heisenberg that his rule for the combination of A and B is that of Cayley, from his 19th century discovery of matrix algebra.*]

The noncommutativity of A and B was taken, according to the Copenhagen School, to mean that A and B are not simultaneously measurable qualities of matter. In the exceptional cases where there is an observable O such that $AO = OA$, one must say that A and O are simultaneously specifiable. Or, to put this another way, if the measurable properties of micromatter represented by A and O are simultaneously specifiable, then their matrix expressions must commute, i.e. $AO = OA$.

If H represents the measure of the energy of a microelement of matter (called the "Hamiltonian") and if C is the measure of any other physical property, then Heisenberg derived the equation for the commutator of the matrices of H and C as follows: $(H, C)_{nm} \equiv (HC - CH)_{nm} = (ih/2\pi)(\partial C/\partial t)_{nm}$. This is "Heisenberg's equation of motion." (*The time derivative $\partial/\partial t$ applies to all of the matrix elements C_{nm}.*)

This is to be compared with Schrödinger's wave equation: $H\psi_n = (ih/2\pi)(\partial \psi_n/\partial t)$. In the Heisenberg formulation, it is the observable

property of the matrix C that changes in time by virtue of the attempt to measure H and C simultaneously, while keeping the states of the matter unchanged. In Schrödinger's formulation, the measurement of the energy of the microelement of matter in a particular state ψ_n changes this state of the matter in time. [*The noncommutativity of the matrices in Heisenberg's formulation corresponds to the noncommutativity of the 'operators' (that determine the wave functions in particular states) in Schrödinger's formulation of wave mechanics.*]

The Heisenberg Uncertainty Principle[43]

An important example of Heisenberg's noncommutation rule applied to measurements is in regard to the measure of the position of a particle in the direction x, and the measure of the momentum of this particle in same direction. In this case the commutator in question is: $(x, p)_{nn} \equiv (xp - px)_{nn} = ih/2\pi$, where $i = (-1)^{1/2}$ and h is Planck's constant. According to Heisenberg's interpretation, this means that the position and the momentum of the particle are not simultaneously specifiable. This leads to the *Heisenberg uncertainty principle* as follows: If $\Delta p = [p^2_{av} - (p_{av})^2]^{1/2}$ is the root-mean square of the measured value of the momentum of the particle, — which we call its uncertainty — and $\Delta x = [x^2_{av} - (x_{av})^2]^{1/2}$ is the root-mean square of the measured value of its position, — its uncertainty — then Heisenberg showed from his commutator of x and p, (x, p), that these 'uncertainties' of measurement satisfy the inequality relation $\Delta p \Delta x \geq h/4\pi$.

Noting the de Broglie relation between momentum and wavelength, $\Delta p = h\Delta k$, where k is the 'wave number' $k = 2\pi/\lambda$, where λ is the de Broglie wavelength of the particle, the Heisenberg uncertainty relation is equivalent to the "Schwarz inequality" for a packet of waves: $\Delta k \Delta x \geq \frac{1}{2}$ (as discovered independently of Heisenberg). Here, Δk is the uncertainty in the measured value of the wave number k of the packet of waves and Δx is the uncertainty of the measure of the position of a component wave. That is, multiplying Schwarz' inequality relation by h, one arrives at Heisenberg's inequality relation.

The derivation of these inequalities depends on the use of the Fourier integral theorem. The latter theorem is only valid in a linear formalism. Thus, a nonlinear theory, such as Einstein's general relativity, would reject the Heisenberg uncertainty relations. Yet, it has been claimed by physicists that the 'uncertainty principle' is a general law of nature, under all conditions. This is clearly false. The uncertainty relations are also based on a model of matter that is atomistic. It is out of context in regard to the holistic, continuous field theory implied by Einstein's theory of matter in general relativity.

The Subjectivity of Matter in the Copenhagen View

What we have discussed so far is that the Copenhagen interpretation of quantum mechanics *defines* matter in terms of the measurements of the physical properties of micromatter by a macroapparatus. The macrodomain is represented with classical variables and the microdomain is represented with quantum mechanical variables. What distinguishes the macrovariables from the microvariables is that the former entails quantities of classical action that are large compared with the quantum of action that is Planck's constant h.

Thus, with the Copenhagen view, at the outset we have a *closed system* $[M|\psi]$, where M is the classical measuring apparatus associated with the property of this matter, and ψ is the variable of the micromatter whose properties are to be measured. What we can say, quantitatively, about the measured matter depends on where we place the vertical line between M and ψ. Suppose that we move this line while still maintaining that the action of the measuring apparatus is large compared with Planck's constant. Moving the vertical line by an arbitrary amount, the closed system then changes to $[N|\varphi]$, where N is the new *observer* of the measured property and φ corresponds to the state of the remaining part of the closed system that is the micromatter *observed*. Thus, the predictions from the measurement N of φ would be different than the predictions made by M for *the same matter*, represented by ψ. This is then a subjective definition of the micromatter — its physical properties, such as inertial mass and intrinsic energy, depends on our mode of

macroscopic measurement. In general, M and N are not simultaneously measurable properties of matter, thus, $MN - NM \neq 0$.

Because of the linear superposition principle, any solution $\varphi_m = \Sigma_n a_{mn} \psi_n$, where a_{mn} is a non-zero coefficient of the probability amplitude ψ_n, among all of the possible probability amplitudes that define the measurement — in the ψ mode of description of the observed. Thus we see that the variability of the location of the vertical line implies *subjectivity* in the *definition* of the properties of elementary matter. That is to say, the physical properties of micromatter depend on the way in which the measurement is carried out by a macroobserver.

In the foregoing example, the equation that determines the property associated with M, when the matter is in the state ψ_n, is $M\psi_n = m_n \psi_n$, where the measured value is m_n. (This is called an "eigenfunction equation"; m_n is an "eigenvalue" of the operator M.) Similarly, the property N, when the particle is in the state φ_m, is determined by the eigenfunction equation $N\varphi_m = n_m \varphi_m = n_m \Sigma_n a_{mn} \psi_n$. We see the subjectivity here in that the 'eigenvalues,' the observables $m_n \neq n_n$, depend on the mode of measurement by the observer.

The Principle of Complementarity[44]

Associated with the subjectivity of the physical qualities of the elements of matter is the primary philosophical basis of quantum mechanics — *Bohr's principle of complementarity*. This is a generalization of the concept of wave–particle dualism, as we discussed earlier. In this view, *truth is pluralistic*. That is, there are many dichotomous truths that are mutually acceptable, even though they are logically inconsistent with each other — as long as these truths are determined separately, under different experimental conditions.

In contrast with this philosophy, the theory of relativity is *monistic*, as we will discuss in a later lecture. It entails a single explanatory base for all of the truths of the natural world, in a logically consistent form.

Einstein's Photon Box Thought Experiment[45]

Einstein proposed a 'thought experiment' to refute the claim of the uncertainty principle. He referred to the relation between energy and time measures, $\Delta E \Delta t \geq h/2\pi$. This relation is derived from Heisenberg's inequality relation between momentum and position. If a particle's only mode of energy is kinetic (i.e. due to its motion) then $E = p^2/2m$. Thus, with the rules of calculus, $\Delta E = 2p\Delta p/2m = v\Delta p = (\Delta x/\Delta t)\Delta p$. Thus, $\Delta E \Delta t = \Delta p \Delta x \geq h/2\pi$.

Einstein's thought experiment is as follows: A box is hanging from the ceiling, connected to a spring scale to measure its weight. The box is heated to a particular temperature, maintained as a blackbody — the 'photons' that are being emitted and re-absorbed by its walls, at equal rates, are in thermodynamic equilibrium with the material of the walls of the box. In this state, the intrinsic energy of the box, E, is maintained at a constant value. Now insert a small door in one of the walls of the box. When a photon approaches this door, a demon opens it to let the photon out of the box, and then it closes it again. The intrinsic energy of the box then decreases because of the decrease of the mass Δm of the box, due to its lost intrinsic energy. (The relation of the intrinsic energy to the inertial mass will be discussed in the next lecture.) The weight reading of the box will then correspondingly decrease by the amount $(\Delta m)g$ and its intrinsic potential energy by $\Delta E = ((\Delta m)g)H$, where H is the height of the box from the floor. The more time that the little door is kept open, the more photons will leave and the uncertainty in the energy of the box will correspondingly increase. Thus, in this case, the uncertainty in the energy of the box, reflected in its weight reading, will increase with the uncertainty of the time during which the measurement of the weight of the box is read (the time taken for the door to be open). That is, ΔE is directly proportional to Δt, rather than being proportional to $1/\Delta t$. This result then refuted Heisenberg's claim.

Bohr responded to this argument by evoking Einstein's theory of general relativity. He said that in the use of the weight of the photon box, one must see this (according to the theory of general relativity) equivalently in terms of a freely falling body in a gravitational potential. Using this argument, Bohr then showed that indeed ΔE is proportional to $1/\Delta t$, in accordance with Heisenberg's uncertainty principle. [*In Bohr' argument, he inserted the relation* $\Delta p \Delta x \geq h/2\pi$ *in order to arrive at the relation* $\Delta E \Delta t \geq h/2\pi$*. But the latter is equivalent to the former, as shown above. Thus his argument was circular — he put into it what he wanted to derive from it. In my judgment, it is therefore invalid.*]

The Einstein–Podolsky–Rosen Thought Experiment[46]

In 1935, Einstein and his collaborators, Boris Podolsky (1896–1966) and Nathan Rosen (1909–1995), presented an argument to show that quantum mechanics is incomplete as a theory of matter. (This 'thought experiment' will be referred to as EPR.) Their argument was as follows: (I will use the model of coupled spins rather than correlated momenta, as EPR did.) Consider two spin one-half electrons, oppositely oriented with total angular momentum equal to zero, binding the two protons of a Hydrogen molecule, H_2.

Now dissociate the two electrons from the molecule, with a spin-independent force, taking one of them to Haifa and the other to Los Angeles. We then experimentally observe the electron in Haifa. We may make a measurement of its angular momentum (spin) value with certainty. Because of the uncertainty relations between the spin and orientation of the electron (1), $\Delta S_1 \Delta \varphi_1 \geq h/2\pi$; if we measure S_1 with certainty then $\Delta S_1 = 0$ and $\Delta \varphi_1 = \infty$. (The latter means that the spinning electron could have any orientation.) The electron (1) could then have spin 'up,' precisely, relative to the direction of an applied magnetic field. Because of the law of conservation of angular momentum, and the fact that initially in the hydrogen molecule the total angular momentum of the two electrons was zero, this means that electron (2), in Los Angeles, must have its spin 'down' relative to this same orientation. Thus, the spin of electron (2) in Los Angeles was determined, without in any way making a measurement on electron (2). EPR then concluded that there must be a complete description of electron (2) independent of making any measurement on it! Reversing this by observing electron (2) in Los Angeles, would then determine the properties of electron (1) in Haifa, without making any macro-measurements on electron (1). Thus they concluded that there must be a complete description of the particles of matter, but that quantum mechanics does not provide it. If one should insist that quantum mechanics is the only fundamental description of elementary matter, then the conclusion is reached that quantum mechanics is both complete and incomplete — a paradox! It is sometimes called the *EPR paradox*.

Thus, Einstein concluded that while the quantum theory does give an accurate description of elementary matter, it is necessarily an incomplete description in the same sense that statistical mechanics is an incomplete description of an ensemble of 10^{24} particles of a mole of a gas. This was the same conclusion that Boltzmann came to in the 19th century, in his analyses of gases. But Boltzmann acknowledged that it is Newtonian physics that underlies the *objective* dynamics of a real gas. The statistics were only used in a *subjective sense* to ascertain the average values of properties of a large ensemble of atoms. On the other hand, the Copenhagen School claims that the probability calculus of quantum mechanics itself is the fundamental law of matter. That is, it is the view here that the laws of nature are laws of chance!

Bohr's response to EPR[47] was that quantum mechanics is as complete a theory of elementary matter that there can be. He said that EPR was misinterpreting the Copenhagen meaning of quantum mechanics. It was Bohr's contention that this theory does not deal with the history of an elementary particle system, say from before two electrons were bound to a Hydrogen molecule to afterwards, when they were unbound. The theory refers only to a measurement, when it is carried out. The uncertainty relations then apply, in the above example, to the separate measurements on electron (1) in Haifa and on electron (2) in Los Angeles. It is independent of the fact that they were previously bound together in a Hydrogen molecule. Thus Bohr concluded that the EPR argument is fallacious because it is *out of context*.

Hidden Variables

In spite of Bohr's reply to EPR, some in the physics community tried to complete the quantum mechanical description of elementary matter. They wanted to satisfy the EPR criticism by adding 'hidden variables' that would provide an underlying deterministic theory for the microscopic elements of matter. An early adherent of this view was Louis de Broglie[48] (his double solution resolution, implying a second *hidden, deterministic* path of a particle of matter) and the later studies of David Bohm (1917–1992).[49]

In Bohm's view, instead of the independent variables, (r, t) of the wave function $\psi(r, t)$, one has an extra independent variable Λ, so that $\psi(r, t, \Lambda) \rightarrow \psi(r, \Lambda(t))$, where $\Lambda(t)$ are the "hidden variables" that trace a trajectory of the particle of matter, *deterministically*. This is analogous to $r(t)$ tracing the trajectory of particle in Newtonian physics. The equation in ψ then becomes more complicated than the Schrödinger wave equation.

One of the troubles with the hidden variable scheme is that if one has an n-body system, the wave function *for a given particle* becomes $\psi(r, \Lambda_1(t), \Lambda_2(t), \ldots, \Lambda_n(t))$, where r is the location of the single particle. Thus the wave function for a particle, say an electron, depends on the trajectories of all of the other particles of the system. This is then called a nonlocal quantum mechanical theory of the single electron.

In addition to this trouble, the mathematician John von Neumann (1903–1957) showed that so long as one stays with the Hilbert space for the probability calculus of quantum mechanics, one may *transform away* the hidden variables — thereby returning to the original formalism![50]

Adherents of the hidden variable approach, such as David Bohm and Jean-Paul Vigier, countered that they wanted to reformulate quantum mechanics without the intrinsic Hilbert space formalism. But this would mean an abandonment of the probability interpretation of this theory. The hidden variable theory and Bohm's extension of it beyond the Hilbert space formulation are still particle theories of elementary matter. They are not compatible with the holistic, continuum, deterministic theory of matter according to the field concept of Einstein's theory of relativity. It is primarily for this reason that Einstein did not accept the hidden variable resolution of the problem of quantum mechanics.

In the next two lectures, we will discuss the basis of Einstein's theory of relativity (special and general) — its implications of new meanings of space, time and motion, and the meaning of the derived relation between mass and energy, $E = mc^2$. Also, we will discuss some the paradoxes inherent in the theory of relativity when it is not properly interpreted. At the end of the course, we will spell out the fundamental incompatibilities between the quantum and relativity theories. These are irresolvable differences, logically, epistemologically and mathematically.

Lecture VII

CONCEPTS OF THE THEORY OF RELATIVITY

The Principle of Relativity

In addition to the quantum theory, a second revolution of 20th century physics was Einstein's theory of relativity. From the conceptual point of view, his earlier discovery of special relativity was not really as revolutionary as his discovery of the theory of general relativity. The former theory initially took an operational, instrumentalist and positivistic view of the laws of nature, not that different from Bohr's Copenhagen interpretation of the quantum theory. But when Einstein came to general relativity as a natural generalization of special relativity, he realized a totally different philosophical point of view for special and general relativity — *as a single theory of relativity.*

It has been said by some that there are two theories of relativity: One is special relativity that covers electrodynamics, particle physics and relativistic mechanics. The second is the theory of general relativity — a theory uniquely of the gravitational force. This is a false judgment. What it is that is 'special' or 'general' is not the theory *per se* (there is only one theory), but rather the 'relativity.' Therefore this law of nature should be called the 'theory of general (or special) relativity,' not the 'general (or special) theory of relativity.' Both special and general relativity are based on a single underlying axiom — *the principle of relativity.* This is the assertion that *the laws of nature must be in one-to-one correspondence in all possible reference frames, from the view of any particular observer of the phenomena explained by these laws.* If the reference frames in which the laws are compared are relatively inertial — implying a relative motion that is a constant speed in a straight line — we are discussing special relativity; if they are in arbitrary types of relative motion, we are discussing general relativity. But both are based on the single axiom — *the principle of relativity.*[51] (*In this lecture we will concentrate on the theory of special relativity. In the next lecture we will discuss the theory of general relativity.*)

The principle of relativity is equivalent, in a philosopher's language, to the assertion that the laws of nature must be fully *objective.* The following objection might then be raised: If a law is indeed a law, then it must, *by definition,* be objective. That is, the principle of relativity seems to be a tautology! It would be like saying: 'woman is female.' This is, of course, a true statement, but it is empty because it is not more than the definition of a word! A mere definition is a necessary truth!

Further reflection shows that this principle is not a tautology because it is based on two tacit assumptions that are not necessary truths — i.e. they are in principle scientifically refutable. The first of these assumptions is that there are laws of nature in the first place. It is assumed that for every effect in nature, there is an underlying cause. Indeed, it is the *raison d'etre* of the scientist to search for the cause–effect relations — the laws of nature — that *explain* the effects observed. This *assumption* is called the 'law of total causation.' It is not a necessary truth.

The second tacit assumption is that we can comprehend and express the truths of nature. It is perhaps arrogant of human beings to believe that we can comprehend any part of the truth of the physical universe. But, as scientists, we do have faith that we can indeed comprehend at least a small part of the laws of nature — the history of science attests to this. Yet, this assumption is not a necessary truth. It is based on faith; therefore it is also a *religious truth.* Recall Einstein's comment "It is incomprehensible to me that I can comprehend any of the truths of the universe." With this statement, we can see the awe in which this great scientist responded to the scientific discoveries of the *truths* of the wonders of nature!

Einstein's Discovery of the Theory of Relativity

Albert Einstein discovered the *principle of relativity* when he was 16 years old. [*This principle was already expounded in a more restrictive sense by Galileo, about 300 years earlier.*] Einstein had just learned that light is explained as a manifestation of electromagnetism, expressed with Maxwell's equations. He learned that indeed light is the propagation of a transverse electromagnetic wave (a wave whose vibrations are perpendicular to its direction of propagation) at a speed whose value in a vacuum (its maximum value) is $c \approx 3 \times 10^{10}$ cm/sec.

Einstein then asked this question: What would be the description of light at rest? That is, if one should move at the speed of light next to a propagating light beam, one should be able to see it at rest. This is like the situation of a passenger on a train moving next to another train on a parallel track, at the same speed. The passenger should then see the other train at rest!

Einstein realized that the answer to his question would depend on the determination of a solution of Maxwell's equations for light, described at rest relative to an observer. Thus, the observer would be moving at the speed of light next to a propagating light beam. Einstein was surprised to find that there are no solutions of these equations that describe light in any other state than moving at the speed of light c, relative to *any* observer, even if the observer is also moving at the speed of light! This conclusion seemed to defy common sense!

Then Einstein realized that his conclusion was based on the assumption that Maxwell's equations (the law of nature for light) had to have the same form in all frames of reference. If he would be allowed to change the form of a physical law every time he would change frames of reference, then he may be able to recover the common sense conclusion. But he was not willing to do this! It was, instead, his assertion that if a particular set of mathematical relations is the expression of an *objective* law of nature, it must be unchanged in form in all possible reference frames, from the view of any particular observer — even if it should defy common sense! After all, he might have asked, what is common sense other than the habit of thinking that we develop, as human beings, a habit of thinking that could very well be in error? Thus, in 1895, at the age of 16, Einstein declared the truth of the *principle of relativity* — that *all* of the laws of nature must be in *one-to-one correspondence* in all possible reference frames, from the view of any particular observer. The *principle of relativity* logically underlies the theory of relativity in its special or general form.

In the expression of the laws of nature, the space and time parameters are the 'words' of a language whose sole purpose is to facilitate the expression of the laws of nature. These 'words' are then *relative to* the frame of reference. (This is the reason for the name of this theory.) But the law itself is not relative — it is independent of the frame of reference, according to the *principle of relativity*. This is analogous to verbal languages, where their words and syntax may change from one reference frame (say, one culture) to another, but the meanings of the sentences remain unchanged. For example, a young American man visits Israel and meets a young Israeli woman and they fall in love. The man says to her in English, "I love you." The young lady responds in Hebrew, "Anee ohevet Otcha." They are each expressing a thought with the same meaning, because they feel its truth. But the language of each (in their own frame of reference) is different. If the young man would modify his declaration of love to: "I love you, most of the time" and the young lady responded as before, of her everlasting unaltered love, the meaning would not be preserved, and the principle of relativity would not be in effect — the relationship would not be a true one. This is the idea: for the relationship to be true *the principle of relativity* must be obeyed — a statement of truth — in this case, true love, must be in *one-to-one correspondence* in all reference frames.

The Spacetime Metric in Special Relativity

As we discussed earlier, Einstein's initial discovery was that Maxwell's equations, the law that predicts the propagation of light, implies that its speed (in a vacuum) is independent of any inertial frame of reference in which this light is described, from the view of any observer. Recall that 'inertial frame' refers to reference frames that are in motion at a constant relative speed, in a straight line.

Consider a one-dimensional space (for convenience); the constancy of the speed of light means that $c = \Delta x/\Delta t = \Delta x'/\Delta t'$, where (henceforth) the symbol Δ denotes an *interval*. We see here a departure from Galileo's relativity principle, where the time measure is absolute: $\Delta t = \Delta t'$. In Einstein's theory the time measure, in the expression of a law of nature, as Einstein originally discovered in regard to Maxwell's equations, is dependent on the frame of reference, as is the space measure. In Galileo's relativity, the time measure is independent of the frame of reference. That is, to preserve the form of a law of nature, such as Maxwell's equations to describe light, the space and time measures must transform in a particular way so as to give *scale changes* of space *and* time measures in the moving reference frames. Thus, in special relativity, in one-dimensional space, $\Delta s = c\Delta t - \Delta x = \Delta s' = c\Delta t' - \Delta x' = 0$, for light propagation, where the 'primed' and 'unprimed' coordinates refer to any arbitrary reference frames that move at a constant speed (or are at rest) relative to each other. Δs is called the 'metric' of the spacetime. In three-dimensional space, the (squared) metric is: $\Delta s^2 = c^2\Delta t^2 - (\Delta x^2 + \Delta y^2 + \Delta z^2) = \Delta s'^2 = c^2\Delta t'^2 - (\Delta x'^2 + \Delta y'^2 + \Delta z'^2) = 0$.

Einstein's extension was then that the spacetime metric, Δs, not necessarily for a light signal, is invariant for all interactions between matter. That is to say, for all interactions that are described by a law of matter, the law remains covariant (the same in all inertial reference frames) with respect to the transformations between reference frames that are the same as those that leave $\Delta s = \Delta s' \neq 0$ (or $= 0$ for light propagation).

The Light Cone

The light cone is a geometric diagram with the time change plotted in the y-direction (the 'ordinate') and the spatial change plotted in the x-direction (the "abscissa"). If the metric $\Delta s = c\Delta t - \Delta x$ is a positive number, then (in the case of one-dimensional space) $c\Delta t$ is greater than Δx. In this case, spatial points are close enough that the signal (the transferred interaction between matter at each end of the spatial interval Δx) has enough time to traverse this spatial separation. These intervals are called "timelike." On the other hand, if the metric Δs is a negative number, then $c\Delta t$ is less than Δx. In this case, the matter at each end of the spatial interval Δx is too far apart for the 'signal' to go from one end to the other in the time specified. These intervals are called "spacelike." The lines connecting the apex of the cone (the origin — the 'present') to points within the cone are "timelike" — denoting causal interactions. They indicate predictions (effects in the future, coming from the present) or retrodictions (effects at the present, coming from the past). The lines connecting the present with points outside of the light cone are "spacelike" — indicating no interaction between matter at these separated places. The lines connecting the present with points on the surface of the light cone denote the propagation of light itself. [*If we should take* $c = \infty$, all *points in the spacetime domain would be connected. This is 'action at a distance' — where all interactions are spontaneous, no matter how far apart interacting things may be! It is at the basis of Newton's theory of gravity. In contrast, the gravitational force, in relativity, propagates at the finite speed c between interacting matter.*]

Lorentz Transformations

At the basis of the theory of special relativity (the principle of relativity applied to inertial frames of reference) are the *scale changes* of the space and time measures in the laws of nature, to preserve their forms in the different reference frames. The Galilean transformations, $\Delta x' = \Delta x + V\Delta t$, $\Delta t' = \Delta t$ do not maintain the invariance of the metric $\Delta s^2 = c^2\Delta t^2 - \Delta x^2 = c^2\Delta t'^2 - \Delta x'^2$, where V is the constant relative speed between the reference frames. But the following transformations do maintain this invariance:

$$\Delta x' = (\Delta x + V\Delta t)/[1 - (V/c)^2]^{1/2}, \quad \Delta t' = [\Delta t + (V/c^2)\Delta x]/[1 - (V/c)^2]^{1/2}.$$

These are the "Lorentz transformations." They are the changes of coordinates that were found to maintain the form of Maxwell's equations in relatively moving inertial frames of reference. [Note that if we would have *action at a distance*, $c \to \infty$, $\Delta x' \to \Delta x + V\Delta t$, $\Delta t' \to \Delta t$. These are the transformations defined by Galileo's principle of relativity.]

It was then Einstein's generalization that the invariance, $\Delta s = \Delta s'$, with respect to changes to any inertial frames of reference, applies to *all* of the laws of nature, not only those of electromagnetism (i.e. Maxwell's equations). This is a statement of the theory of special relativity.

Relative Simultaneity

If two events happen simultaneously in the 'unprimed' frame of reference, then $\Delta t = 0$. In this case, the time measure in the 'primed' frame of reference is not zero. It is, according to the formulas in the preceding paragraph: $\Delta t' = (V/c^2)\ \Delta x/[1 - (V/c)^2]^{1/2} \neq 0$. That is, if two events are seen to be simultaneous in one frame of reference, they will not generally be seen to be simultaneous in other frames of reference. Thus, simultaneity, in special relativity theory, is relative to the frame of reference in which interactions are described.

This conclusion should be examined closely in view of our meaning of the space and time measures, not as physical entities, but rather as relative language parameters in the *expression* of a law of nature. Consider the following 'thought experiment': A cat is crossing an intersection in the road. At the center of the intersection there is a manhole cover and at the curb there is a workman with his hand on a lever that would open the manhole cover. As soon as the cat steps toward the manhole cover, the workman *simultaneously* pulls the lever that opens it. The cat then falls below the street, never to reach the other side of the road.

Suppose now that a helicopter is flying over the road, at a speed close to the speed of light! In looking down at the road, the pilot sees the cat crossing the road and the workman pulling the lever to open the manhole cover *non-simultaneously,* i.e. at a different time than when the cat reaches the manhole cover in the road. The pilot then expects the cat to reach the other side of the road. But instead he sees that the cat disappears midway across the road! He then asks himself: "why didn't the cat get to the other side of the road?" He answers that it was because what he saw was influenced by the fact that he was in a moving frame of reference, relative to the cat and the road. To learn what really happened he applies the Lorentz transformation to put himself into the frame of reference of the cat and the manhole cover, independent of any outside observer! This is called the 'proper' frame of reference — it involves only the interacting things — the cat and the Earth that pulls it downwards. In this (proper) reference frame, he learns that the cat did not reach the other side of the road because the workman pulled the lever at the precise time when the cat stepped down toward it, and so it fell below the street before reaching the other side.

Thus we see that the relativity of simultaneity in this theory *is not physical*; it is only *descriptive* regarding a viewing from the frame of reference of the observer. To say that relative simultaneity is a physical fact is to predict a paradox — that, in this example, the cat would reach the other side of the road and it would not reach the other side of the road!

Time Contraction and the Twin Paradox

If the separation of spatial points Δx is zero in one reference frame, then the Lorentz transformation of time measures is: $\Delta t = \Delta t'[1 - (V/c)^2]^{1/2}$. Here, Δt is a time interval in the language of a law of nature expressed by a fixed observer of a physical process in a frame of reference that moves relative to him at the speed V. The time interval $\Delta t'$ is the time measure in the language of the law of nature (by an observer) in the moving frame itself. According to the theory of relativity, then, there must be a change in the *time measure*, in transforming from one reference frame to another. That is, Δt is less than $\Delta t'$. But according to the meaning of the space and time measures, this is only a *scale change*. That is to say, the 'fixed' observer may have to put eight digits on the face of a moving clock to *describe a law of nature* in that frame with the transformed time and space measures. But this does not mean that anything physical has happened to the spring behind the face of the moving clock — by virtue of its motion relative to an observer — that does not happen to the spring of the observer's own clock!

The claim is made by many contemporary physicists that the time change is not a scale change — that it is a real, physical change. If this were the case, a moving clock (or a moving human being) would age at a slower rate than the aging of the fixed clock (or observer). But we must recognize what Galileo discovered — that *motion, per se*, is strictly a subjective aspect of the *expression* of a law of matter. That is, one could just as well call the 'moving body' the 'fixed body,' and vice versa, in the representation of a law of matter. It was Galileo's reflection that while it is true that the Earth moves relative to the Sun, from the Sun's perspective, it is equally true that the Sun moves relative to the Earth, from the Earth's perspective! Thus he did not agree with Copernicus that there is an absolute center of the universe — be it the Sun, the Earth or anything else!

If, for example, the time change in the description of any process in nature is a physical change in aging, then when a twin sister space pilot goes on a round trip journey from her sister at home, she would age less than her sister during the journey, from the perspective of the stay-at-home sister. But from the perspective of the traveling sister, it is the stay-at-home sister who making the round trip journey and the stay-at-home sister would be younger than her pilot sister after the completion of the round trip journey! Thus, with this interpretation, it would have to be concluded that when the traveling sister returns from her round trip, she would be both older and younger than her sister. This is called the "twin paradox."

The error in this conclusion is the faulty interpretation of the time measure as an *objective* physical change instead of a *subjective* scale change, when transforming to different reference frames![52]

The Fitzgerald–Lorentz Contraction

The same argument as above holds for relative spatial measures. If the locations of the ends of a stick are measured at the same time ($\Delta t = 0$) then $\Delta x = \Delta x'[1 - (V/c)^2]^{1/2}$. This means that the observer's measure of the extension of the stick Δx, moving at the speed V relative to the observer, is less than the extension of the stick $\Delta x'$ in the frame of the stick itself. But this is not a *physical measure* of length in the frame of reference that moves relative to a fixed observer. It is only a *scale change* of spatial extension, in the *description* of the laws of nature in the different reference frames. Again, with this interpretation, it does not imply a paradox.[53]

The Transformations of Velocities in Relativity Theory

Defining the velocities in the different inertial reference frames in terms of differential elements instead of finite intervals, dx, dt, dx', dt', then the above Lorentz transformations predict that the velocity of a body in the moving frame according to the observer in the fixed frame is:

$$v' = dx'/dt' = (dx + Vdt)[1 - (V/c)^2]^{1/2}/[1 - (V/c)^2]^{1/2}[dt + (V/c^2)dx]$$
$$= [dx/dt + V]/[1 + Vv/c^2] = [v + V]/(1 + Vv/c^2).$$

In the classical limit (*action at a distance*), $c \to \infty$, so that $v' \to v + V$ (the Galilean transformation of velocities).

In the theory of relativity, if $v = c$, then $v' = (c + V)/(1 + V/c) = c(1 + V/c)/(1 + V/c) = c$.

That is, if a body moves at the speed of light c in any reference frame that moves at the speed V relative to an observer, its velocity, as seen from any fixed reference frame is also equal to c. (That is, in classical physics one would say that $v' = c + V$, in relativity physics, $v' = c$.) This was the result that seemed to Einstein to defy common sense. But it is a conclusion that is compatible with the covariance of the laws of nature — that is, their *one-to-one correspondence* in all relatively moving inertial frames of reference. Einstein rightly concluded that the objectivity of the laws of nature is to be trusted over our human common sense. It was this idea (as Plato taught), on the abstract and rational approach to the truths of nature, that led to the great discoveries of the theory of relativity.

In the next lecture, we will discuss a few more of the paradoxes of the theory of special relativity, when the space and time transformations are interpreted physically, rather than as *scale changes* — the grandfather paradox, interstellar travel, Gödel's paradox. Then we will show the origin and interpretation of the energy mass relation, $E = mc^2$. Finally we will proceed to the ideas of the theory of general relativity.

Lecture VIII

FROM SPECIAL TO GENERAL RELATIVITY

The Paradoxes of Time Travel

The interpretation of the space and time parameters in special relativity theory as physical experiences leads to genuine logical paradoxes. These may be resolved only by re-interpreting them as *measures* in the language of the laws of nature: The transformations of the space and time measures, in turn, from one reference frame to another, in order to preserve the forms of a laws of nature, are only *scale changes*, they are not physical changes.

The following are examples of paradoxes that are encountered when the faulty interpretation is used in applying the theory of relativity.

1. *The grandfather paradox.* This is usually presented in terms of a person traveling backward in time to meet his grandfather, who is physically younger than he is at their meeting. Here we will suppose that a surgeon goes backward in time to meet his father, before his father met his mother. He then forces his (young) father onto a table and operates on him to make him sterile. Thus, this surgeon could never have been conceived and born — *he does not exist*! *But he does exist* — to have taken this

trip into the past. This is a logical paradox that clearly demonstrates the error in interpreting the time measure in relativity theory as a physical experience! It is similar to the error of the *twin paradox*, discussed in the previous lecture.[54] [*The way that some physicists try to get out of the grandfather paradox is to resort to the many-universe idea. They say that the person (the surgeon in this example) exists in one universe, but he does not exist in a different, parallel universe, that he crosses over into in the past! I believe that this is* very far fetched! *It is not based on the logic of real science! I believe that the claim that it is a bona fide scientific 'resolution' of the paradox insults our intelligence! Here, I have used my right as a 'professor' to profess! The students have every right to disagree with me on this, so long as it is based on scientific logic*!]

2. *Interstellar travel.* It is said by some (present-day) scientists that it should be possible for a person to travel, in his lifetime (of, say, 100 years) to a star that is 1000 light-years away. Without the relativity formulas it would take 1000 years to travel this distance, at the speed of light. But if we interpret the time contraction formula (the Lorentz transformation) as a physical experience, then *from the view of the star*, it might take the traveler only a day of his life to reach it. But during his travel time of one day, from the star's reference frame, the star will have used up 1000 years worth of its nuclear fuel. On the other hand, *from the view of the traveler*, who would see the star coming toward him and the earth departing from him, he would age 1000 years during the trip to the star (if he could live that long!), but the star would have used up only one day's worth of its nuclear fuel! If both statements are true, we must come to the paradoxical conclusion that after the traveler reaches his destination, the star will have used up 1000 years worth of its fuel *and* it will have used up one day's worth of its fuel! The error leading to this logical paradox is the faulty assumption that the *scale change* for the time transformation in special relativity (i.e. the Lorentz transformation) is a physical experience. This is clearly an illogical interpretation.

3. *Gödel's paradox.* Kurt Gödel (1906–1978) discovered a *cyclic* geodesic solution of Einstein's field equations in general relativity, for the case of a constant matter density of the universe.[55] [*The ideas of Einstein's general relativity will be discussed later on in this lecture.*] If these geodesic paths in space and time are the physical experiences of matter, say a human being, then a peculiar prediction follows: Let us say a person, Jenny, in traversing a single cycle, meets herself, Jenny′. She tells Jenny′ that people are 'multivalued.' It is in Jenny's memory when she tells Jenny′ that she is herself! Jenny′ has no such memory and she thinks that Jenny is insane! Jenny′ then goes on her way on the same cyclic geodesic. Eventually she meets herself, as Jenny″, and tells her the same story. Jenny″ thinks that Jenny′ is insane and proceeds on the same cyclic geodesic, and so on, *ad infinitum.* Gödel responded to his paradoxical conclusion in saying that the interpretation of the space and time parameters of the language of the theory of relativity should be re-examined. Indeed so! They are not physical experiences. The cyclic geodesic path is a geometrical feature of the spacetime language that is used in the theory of general relativity. It is not the life of a human being!

We saw in the preceding three examples of time travel that the time and space parameters are not to be interpreted as physical experiences; otherwise one gets into the problem of paradoxes. It was not Einstein's intention to use this interpretation when he discovered the principle of relativity. It is the law of nature that predicts physical effects, not the language that we use to express these laws. (*Plato would have agreed with this judgment.*)

The Energy–Mass Relation $E = mc^2$ in Special Relativity

Consider the observation of a radioactive material randomly emitting 'photons' (electromagnetic radiation). The measured energy of the matter, before it emits a photon, is E_0, its measured energy after it emits the photon is E_1 and the energy of the photon is E_γ. According to the law of the *conservation of energy*, $E_0 = E_1 + E_\gamma$.

Suppose now that one observes these events when the radioactive material is in motion at the speed v relative to the observer. In this case, the energies measured would be different than in the former case; they would be: E_0', E_1' and E_γ'. The principle of relativity then dictates that the law of *conservation of energy* must hold in the moving frame, with the same form as in the fixed frame: $E_0' = E_1' + E_\gamma'$.

Subtracting the latter equation from the former, we have:

$$(E_0' - E_0) - (E_1' - E_1) = (E_\gamma' - E_\gamma). \tag{1}$$

The difference between the matter energies E_0' and E_0 is that the primed reference frame corresponds to viewing the emission in a moving frame while there is no relative motion in the unprimed frame. When the velocity of the matter is small compared with the speed of light, this difference must be the classical kinetic energy of the matter, i.e. when $v/c \ll 1$, $(E_0' - E_0) \approx (1/2)(mv^2)_0$. This is because, in accordance with the *correspondence principle*, the predictions of special relativity theory approach the forms of the predictions of classical physics, when $v/c \ll 1$. Similarly, in this limit, $(E_1' - E_1) \approx (1/2)(mv^2)_1$.

From the transformation of Maxwell's equations from one inertial frame to another to preserve its form, it is found that $E_\gamma' = E_\gamma/[1 - (v/c)^2]^{1/2}$. When v/c is small compared with unity, (with a binomial expansion of the denominator) this transformation becomes: $E_\gamma' = (1/2)(v/c)^2 E_\gamma$. Thus, the nonrelativistic limit of Eq. (1) is:

$$(1/2)[(mv^2)_0 - (mv^2)_1] = E_\gamma(1/2)(v/c)^2. \tag{2}$$

On the left hand side of Eq. (2) the velocity v does not change from the time before the photon is emitted to the time after it is emitted — v is constant. Thus, since the right hand side of this equation is a positive number, the left hand side must be a positive number. This must then entail a change of the inertial mass of the emitting body. Thus we have from Eq. (2)

$$(1/2)v^2(m_0 - m_1) = (1/2)v^2\delta m = (1/2)(v/c)^2E_\gamma$$

or,

$$E_\gamma = \delta mc^2,$$

where $\delta m = (m_0 - m_1)$ is the decrease of the mass of the body when it emits the photon with energy E_γ. Thus, if the *total mass* m of the body would be converted into radiation, the *total energy* emitted would be $E = mc^2$. This is the intrinsic energy of the body with inertial mass equal to m. If one would be observing this body from a frame of reference that moves at the speed v relative to an observer, its total energy would be seen to be: $E = mc^2/[1 - (v/c)^2]^{1/2}$. Note that if $v/c \ll 1$, a binomial expansion of the form for E becomes a series whose first two terms are $E = mc^2 + (1/2)mv^2$ — the intrinsic energy of the matter plus its kinetic energy. [*This energy–mass relation may also be derived more rigorously without appeal to radiation emission, in relativistic dynamics, yielding the same result.*]

The magnitude of the intrinsic energy (called "rest energy") of this matter, mc^2, is much greater than the classically measured energy (say the kinetic energy) of the body — that is the order of mv^2. If $v \ll c$, say typically v is the order of 100 cm/sec, then $v/c = 10^{-8}$. In this case the ratio of the rest energy of the body to its kinetic energy would be the order of $(v/c)^{-2} \approx 10^{16}$. That is, the rest energy of a body would be the order of ten thousand million million times greater than the ordinarily measured energy.

One might then ask: Why wasn't this huge rest energy of matter not detected before the early decades of the 20th century? It is because energy, in itself, is not normally measured; it is the difference of energies that is measured (the energy transferred in an interaction between matter). This entails a difference wherein the rest energy cancels, i.e. $\delta E = (E_0 + mc^2) - (E_1 + mc^2) = (E_0 - E_1)$.

However, in the early 20th century, it was discovered that a heavy unstable nucleus, such as an isotope of uranium, with mass M, may fission (break apart into smaller nuclei), releasing particles whose sum of masses $M_1 + M_2 + \cdots$ is not equal to the original mass

M of the unstable uranium nucleus. It was found that the binding energy of the original unstable nucleus is ΔMc^2 — where ΔM is the difference between the mass M of the fissioned nucleus and the sum of parts it released, $M - (M_1 + M_2 + \cdots)$. [*This discovery not only verified Einstein's derivation of the rest energy of matter from special relativity, it also initiated the age of nuclear technology — leading to the application of the peaceful use of nuclear energy and (unfortunately) the invention of the nuclear bomb*!]

The Meaning of $E = mc^2$

It has been said by many physicists and philosophers that the formula $E = mc^2$ means that "mass is equivalent to energy." This is philosophically false. It is not what Einstein said when he derived this relation. What he said was that "the inertial mass of matter is *a measure* of its energy content."

In physics, as Newton originally postulated, the inertial mass of matter is, *by definition*, a measure of its resistance to a change of its state of rest or constant motion. The energy of matter, on the other hand, is *by definition*, the capability of this matter to do work. Thus, mass and energy are totally different concepts! What should be said, instead of saying that mass is equivalent to energy, is that *mass* (the inertia of matter) is a *measure* of the capability of this matter to do work (its intrinsic *energy*). That is, analogously in Hebrew, a *tapooz* (an orange) is not *equivalent to* an *aitz* (a tree) because the orange comes from a tree! Similarly, the speed of a body, v, is not *equivalent to* its kinetic energy, because the kinetic energy depends on the speed v!

The Theory of General Relativity[57]

The covariance of the laws of nature, *the principle of relativity*, in special relativity — their objectivity with respect to transformations to inertial frames of reference, is (1) a too restrictive relativity, and (2) an ideal (and unachievable) limit. A test body, anywhere, is subject to forces exerted on it, by other matter that is *anywhere* in the universe. This force causes the test body to accelerate relative to the position of the force-exerting matter. It is only in the ideal limit of a totally vacuous universe that the reference frame of the test body would be truly inertial!

Thus we see that the theory of special relativity is not a valid truth, except in an ideal limit, when the universe, or any portion of it, may be *approximated* by a perfect vacuum, *everywhere*. *The principle of relativity*, applied to the real world, then asserts that the laws of nature are in *one-to-one correspondence* in all possible *non-inertial* frames of reference, from any observer's view. *This is the assertion of the theory of general relativity.*

The Metric of a Curved Spacetime

Since the distribution of the matter of the universe is *continuously* variable, the metric of spacetime must be correspondingly continuously variable. The metric of the spacetime in special relativity, as we have seen earlier, is (in differential form) $ds^2 = c^2dt^2 - dr^2$, where $dr^2 = dx_1^2 + dx_2^2 + dx_3^2$ is the differential three-dimensional spatial element. The coefficients of the space and time elements of ds^2 are (1, −1, −1, −1), they are not variable. This metric describes a *flat spacetime*. The family of its geodesics is a family of straight line paths. *It is the rule in special relativity theory that all of the laws of nature must maintain their forms, in one-to-one correspondence, under the same spacetime transformations that keep* ds^2 invariant, *i.e.* $ds^2 = ds'^2$.

Einstein then generalized this geometrical feature of special relativity to the non-Euclidean geometry, characterized by a Riemannian metric as follows: $ds_{\mathrm{gr}}^2 = g_{00}(x)dx_0^2 + g_{01}(x)dx_0dx_1 + \cdots + g_{33}(x)dx_3^2 = \Sigma_{\mu\nu}g_{\mu\nu}(x)dx_\mu dx_\nu$. The coefficients of the differential elements, g_{00}, g_{01}, ... are the ten fields, $g_{\mu\nu} = g_{\nu\mu}$ (the "metric tensor"), dependent on the space and time coordinates x, (μ, $\nu = 0, 1, 2, 3$). They are the 'signature' that replaces the constants (1, −1, −1, −1) of special relativity. *The rule of general relativity theory is that the forms of the laws of nature must remain in one-to-one correspondence under the same transformations that leave the generalized metric invariant, i.e.* $ds_{\mathrm{gr}}^2 = ds_{\mathrm{gr}}'^2$. This metric of the spacetime predicts that the corresponding geodesics are not straight lines — *they are continuously variable curves.*

A "geodesic" is a continuous path such that the distance between any two of its points is a minimum (or, in principle, a maximum). For a test body to move off of a curved geodesic would require external energy (work) to be applied to it. In special relativity, Newton's second law of motion predicts that an external force would cause the body to accelerate, i.e. to depart from its straight line motion. In general relativity, similarly, it would take external energy to move a test body off of its curved geodesic path.

The family of these continuous curved geodesics is called a "curved spacetime." The prediction is that a test body in such a spacetime will move naturally along a geodesic curve. Thus, the acceleration of matter is a built-in feature of this geometric generalization.

There are features of the curved spacetime that are not visualizable with our senses. For example, in Euclidean space, if one should look at the surface of a hollow sphere from the inside, one would see it as a concave surface (positive curvature). Looking at the sphere from the outside, it would appear as a convex surface (negative curvature). In contrast, in Riemannian space, the curvature of spacetime, has one sign, as viewed from anywhere.

The Principle of Equivalence[58]

In accordance with what has been said thus far, if a material body, such as a planet, is moving 'freely' on a curved geodesic in a Riemannian spacetime, with the same orbital trajectory about the sun that is predicted by a gravitational force of the sun in a Euclidean spacetime, then the geometrical description of the body's 'free' trajectory in the curved spacetime is *equivalent to* the application of an external force on that body in a Euclidean spacetime. This is a general expression of *the principle of equivalence.*

The program of the theory of general relativity is to determine, *from first principles*, the explicit relation between the material content of a physical system and the geometrical equivalent of this system (in any domain, from elementary particles to cosmology). Einstein's field equations were so derived. They have the (symbolic) form: G(geometry) = T(matter). That is, given the matter variables on the right, the geometrical variables (that entail the metric tensor solutions $g_{\mu\nu}$) on the left are determined from the solutions of these equations. (*These are ten second order, nonlinear differential equations*). Once these ten solutions [the components of the symmetric metric tensor $g_{\mu\nu}(x)$] are found, they may be inserted into the geodesic equation, to determine the motion of matter in the curved spacetime. In view of our interpretation of the Einstein field equations as identities, given the metric tensor solutions on the left, the matter field solutions on the right are determined by inverting these equations. [*This interpretation implies that there are no meaningful vacuum solutions, in an exact sense, of the Einstein field equations in which the right hand side of these equations is zero. The vacuum equations in general relativity have yielded curved-space solutions that correspond to empirically tested consequences. But these must be interpreted as approximations for the solutions of the exact form of the field equations, where there is no vacuum (i.e. where there is no zero on the right). This is because, in this view, the existence of matter, anywhere, is represented by a curved spacetime, and the vacuum — a lack of matter in the universe — must, consequently, be equivalent to a flat spacetime.*]

[Physicists tend to finalize an idea with the pronouncement of a "principle" — *the principle of correspondence, the principle of complementarity, the principle of linear superposition, the Heisenberg uncertainty principle, the principle of equivalence, the principle of minimal coupling in electromagnetism.* (This is the principle that asserts that the description of electromagnetic coupling that we have at the present, in terms of the standard Maxwell variables, is all that there can ever be in the explanation of electromagnetism!) *There is no scientific reason for these principles to be absolute truths of nature! To finalize these principles is indeed in itself a principle of dogmatism!*]

The Tests of General Relativity[59]

Thus we see that in general relativity, the geometry of spacetime and its implicit family of curved geodesics, is a way of expressing the physical properties of the constituent matter of the universe — in any of its domains, from that of electron and proton and all other elementary matter to the physics of the universe as a whole, the subject of cosmology.

One of the successes of Einstein's theory of general relativity was its prediction of all of the features of the force of gravity. This was previously explained with Newton's theory of universal gravitation. Einstein's theory, totally different than that of Newton, made the same predictions as the classical theory, in addition to new successful predictions that were not even in principle predicted by the classical theory of gravity.

One of the early tests of general relativity entails observations in our solar system. It was the prediction that a beam of light in the universe would bend as it glances past the rim of the sun. This is not because the sun exerts a Newtonian-like gravitational pull on the (massless) photons of this light beam! — as many have asserted. It is because, according to Einstein's theory, the light beam moves *naturally* along a curved geodesic that is a representation of the existence of the sun.

The bending of starlight, as it propagates past the rim of the sun, was observed, in qualitative and quantitative agreement with Einstein's theory, by a Cambridge University group, headed by A. S. Eddington, (1882–1944) in 1919. It is a prediction of the theory of general relativity not predicted by Newton's theory of universal gravitation. It led to the replacement of the classical theory of gravity that was believed to be true for the preceding 300 years, with the field theory of Einstein's general relativity.

A second crucial test of general relativity was as follows: Because the field of a varying gravitational potential (in the Newtonian sense) corresponds in Einstein's theory to a change of the time measure (according to the *equivalence principle*) the prediction is made that by increasing the gravitational potential in which a gamma ray (electromagnetic radiation of a given frequency) is emitted by a radioactive emitting substance, this measured frequency must correspondingly change to smaller values (*a gravitational redshift*). This effect was observed, both qualitatively and quantitatively, in the 1950s, in agreement with the prediction of the theory of general relativity.[60]

A third crucial test of general relativity was an effect already observed in the 19th century. It is the perihelion precession of the orbit of a planet about the sun. The basic reason for this effect, according to the theory of general relativity, is that functions of space and time may not be separated in the description of the dynamics of a body as they are in classical physics. When they may be separated, as in Newton's theory, where the time measure is absolute, it predicts stationary orbits — that a planet will move on an elliptical orbit *cyclically*, returning to the same place in the sky, relative to the position of the gravitationally attracting sun, in equal times. The astronomical observation was to focus on Mercury at its perihelion point — this is the point in its elliptical orbit that is closest to the sun, whose center of mass is located at one of the foci of the planet's elliptical path. The prediction of Einstein's field equations, that the orbit of a planet is non-stationary and non-cyclic (the perihelion precession of Mercury's orbit) was in close agreement with the observations, both qualitatively and quantitatively.

With the foregoing three crucial tests of Einstein's theory of general relativity, his approach was able to supersede Newton's theory of universal gravitation that was believed to be true for the preceding 300 years. It differs conceptually from the classical theory in two major respects: (1) *action at a distance* is replaced by the propagation of forces (of any type) between interacting matter, at a finite speed, whose maximum value is the speed of light in a vacuum. (This is the order of 3×10^{10} cm/sec.) The second difference is that Newton's approach is based on the model of matter in terms of *atomism*. In Einstein's field theory, on the other hand, atomism is replaced with the model in which there is a closed system — the universe — that is a continuous whole. What appear to be the singular atoms of matter, planets, galaxies, human beings, ... are instead distinguishable manifestations (modes) of this single continuum. Thus, this is a theory of matter based on *holism*, where there are no separable, singular things. There are only correlated modes of a continuous whole. In philosophical terms, this is similar to the approach of Spinoza in the 17th century.

A Unified Field Theory

It was originally Faraday's idea, in the 19th century, that there is a single, general field of force that manifests itself in different ways, depending on the physical conditions. He (and his predecessor H. C. Oersted (1777–1851) showed this, theoretically and empirically, in the unification of electricity and magnetism, into a single field of electromagnetism. It was then demonstrated by Maxwell, in his mathematical formulation of the field theory in the form of Maxwell's field equations, that the physical manifestations of electromagnetism also include all of the physical manifestations of optical phenomena.

Einstein was the primary physicist who tried to extend Faraday's program of a unified field theory, in the 20th century, by extending the theory of general relativity to include electromagnetism with gravity. Another of the great physicists who followed Einstein's lead in this approach was Erwin Schrödinger.[61] Unfortunately they were not successful in their day in this attempt. Nevertheless, in believing the logical basis of the theory of general relativity, as a fundamental theory of matter, both Einstein and Schrödinger knew that it would be demonstrated one day as a natural outcome. Further, Einstein also believed that the formal expression of the quantum theory would appear as a natural inclusion of the field theory of general relativity, not based on the premises and philosophy of the Copenhagen School, but rather based on the premises and philosophy of the theory of relativity. *In the last lecture of this series, we will detail the fundamental differences between the relativity and quantum theories, as fundamental truths of matter.*

Coming back to Faraday's time, he tried to show in experimentation that the gravitational field of force is unified with electromagnetism and optics. He was not able to demonstrate this, empirically, because of the very small magnitude of gravitational coupling compared with the magnitude of electromagnetic coupling of electrically charged matter, in the domains that he studied.

Another problem was the following: The gravitational force, as it was known in the 19th century, is only attractive. That is, matter was never known to gravitationally repel other matter. (*An example of the repulsion of matter from other matter that was discovered in the 20th century was Hubble's discovery of the expansion of the universe. The galaxies are seen to move away from all other galaxies.*) On the other hand, electromagnetic forces are seen to be either attractive or repulsive. This difference would have to be explained in a truly unified theory of electromagnetism and gravity.

Further, in 20th century physics, the forces in the nuclear domain, the 'nuclear forces,' are seen in experimentation to have both attractive and repulsive components. These are the 'strong nuclear forces' that bind neutrons and protons together in the nuclei of atoms, and the 'weak interactions,' that are responsible for the beta decay of radioactive matter (its emission of electrons). These are the claimed four fundamental forces — the (long range) electromagnetic and gravitational forces and the (short range) nuclear (strong) and weak forces, that any unified field theory must eventually explain in a single field formalism.

There is one more feature of matter that must not be forgotten in a unified field theory. It is the inertial manifestation of matter. This is the resistance of matter to a change of its state of rest or constant motion. It has been my contention, in my research program, that the key to achieving a unified field theory is the proper incorporation of the inertial manifestation of matter, in a rigorous way, within the field theory of general relativity, along with the forces exerted by matter on matter. The latter deals with the *action* of matter on matter. The former deals with the *reaction* of matter to the forces exerted on it in the full description of a closed system. Both manifestations must be incorporated in a bona fide unified field theory.

In the next lecture, we will discuss the subjects of astrophysics and cosmology. Then, in the final lecture, we will discuss the basic premises and philosophies of the quantum and relativity theories, side by side, to see where they have fundamental differences, and what may be done to resolve them.

Lecture IX

THE UNIVERSE

We interpret the theory of general relativity as a general theory of matter, in all domains. In this lecture we will discuss (in a strictly non-mathematical fashion) some of its consequences in the physical universe. First, there will be discussion of some of the 'exotic' stars (such as black holes, pulsars and quasars), and dark matter. Then we will proceed to a discussion of the physics of the universe as a whole — the subject of cosmology — in the conventional views as well as a new approach.

Astrophysics

Black Holes

We have seen that in general relativity, the curved spacetime and its constituent geodesics are a geometrical representation for the existence of matter. The more dense that matter may be, the more curved are the corresponding geodesics. It is possible, in principle, then, that a star could be so dense that its geodesics are so curved that they close on themselves. That is to say, all of the geodesic paths that leave such star must return to it. This field of geodesics would be similar to the closed (irrotational) field of the magnetic lines of force of a bar magnet. Since light (or any other signal emanating from matter) must propagate along its corresponding geodesic, any light that leaves this sort of star, must return to it. Any other type of interaction that propagates from such a star must also return to it, i.e. *this type of star could not interact with any other matter*. Such a star would then be 'black' to any outside observer. It is called a 'black hole.' Of course, most stars, if they are imploding in the course of increasing their density would reach a critical point, before the black hole state, when they would break apart. But in rare cases, the physical situation (the internal forces) may be that such that they would not break apart before reaching this state.

There is another type of black hole discussed in today's literature. It is described by the 'vacuum solution' of Einstein's field equations.[62] This solution entails a singularity at a finite distance from the center of the star (called 'the event horizon'). When the size of the star is such that its radius coincides with the distance from the center of the star to this singular surface, the metric becomes infinite and the star would not transmit energy to the outside world. [*In my view, there is a genuine criticism of this type of 'black hole' that entails a mathematical singularity. There is indeed a bona fide reason to exclude all singularities from the acceptable solutions of the theory of general relativity, as Einstein continually professed. In my view, it is based on the laws of conservation of energy, momentum and angular momentum, that require the solutions of the field laws to be analytic and nonsingular everywhere. This is in accordance with Noether's theorem. There is also the criticism that the vacuum equations are not realistic, since, as we have discussed in an earlier lecture, according to the theory of general relativity, the (non-vacuous) matter variables are another way of discussing the geometrical variables of the system, and vice versa. With this interpretation, the Einstein field equations are identities. It is then illogical to consider the vacuum equation (where there is no matter anywhere, thus a null field on the right hand side) as exact. The only solution of the vacuum equations, as an exact set of field equations, is the set of constants, (1, −1, −1, −1), that is, the Lorentz flat space metric. If the vacuum equations predict a finite curvature that gives correct empirical results, then they must be an approximation for Einstein's field equations in their exact form, where they do represent matter.*]

The existence of the first type of black hole mentioned above is, in principle, possible in the context of the theory of general relativity. From the theoretical side, to establish the existence of such a 'black hole' it would be necessary to prove that there are solutions of Einstein's equations for closed geodesics that are stable. [*Early studies indicate that there are no stable solutions for a family of closed geodesics — thus, that black holes do not exist! But these results are only preliminary at this stage.*]

From the empirical side, there are some astronomical studies that imply the existence of a black hole star. One of these is based on the observation of an X-ray emitting star, indicating that some of its emission is missing! It was then postulated that the observed star is bound, gravitationally, to a black hole that is absorbing part of the X-ray emission, which in turn becomes invisible to any outside observer. This is an unacceptable conclusion since the black hole, *by definition*, cannot be bound to another star, other sort of matter, or even to another black hole! If it would be so-bound, then the geodesics associated with the black hole would not return to it; they would have to end at the other matter that interacts with it!

It is often claimed that black holes exist at the centers of galaxies. There does seem to be astronomical evidence that there are very massive and dense stars at these centers. But we do not know that these are black holes! They could be clusters of very dense stars, such as neutron stars, but not as dense as black holes.

There is no conclusive evidence at this point in time, either theoretical or empirical, that black holes exist in nature. Though I am willing to be shown that I am wrong, I remain skeptical! At this stage, it seems to me similar to the claims (from mythology) of several centuries ago, that unicorns or mermaids must exist! — based on physical evidence that could be explained in many other ways.

Pulsars

A pulsar is a very dense, small star that emits pulsating radiation at fixed intervals of time. The standard explanation for this phenomenon by the astrophysicists is that this is a neutron star (a star composed only of bound neutrons) that is emitting radiation as it rotates. The analysis of such a rotating neutron star reveals that it would be seen to emit radiation in periodic pulses. (This would be similar to the observation of an emission from a lighthouse whose light source is in rotation.)[63]

A model of a pulsar that I have considered in my research is as follows: The internal dynamics of a star, as a plasma (a sea of positively and negatively charged matter) indicates that its volume pulsates, increasing and decreasing in periodic fashion. Suppose, then, that a very dense star, such as a neutron star, in the part of its cyclic pulsation that decreases its volume (increases its density) goes into the density state of a black hole, and then in the next part of this pulsation cycle, where its volume increases (its density decreases) it goes out of the black hole state. When it would be in the black hole state, it would not emit radiation to the outside world, and when it is out of the black hole state it would emit radiation.[64] Thus, the observer would see this neutron star emit pulses of radiation with a fixed period. In my theoretical studies, I have found, based on the field formalism of general relativity, that the intensity of these pulses attenuate — which is also empirically true. But the mathematical details of this model of the pulsar must still be worked out. Theoretically, it still depends on the existence of the black hole state, in the first place, which I have indicated that I doubt!

Dark Matter

There has been a great deal of discussion in recent years of the subject of 'dark matter.' Its existence was originally postulated by the astrophysicists to explain the observed rotations of the galaxies.[65]

Most of the galaxies of the universe have a two-dimensional pancake shape, bulging at the center of its plane, where most of the stars are located. The galaxies usually have spiral arms. (Our own Sun is an average sized star, located in one of the spiral arms of our galaxy, *Milky Way*.) The galaxies, including our own, are seen to be rotating about an axis that is perpendicular to its plane.

The questions then arise: Why do the galaxies have the shapes that they do? And why do they rotate at the rates that the astronomers measure?

To answer the second question the astronomers first postulated that the neighboring galaxies to a given one, such as our neighboring galaxy, *Andromeda*, gravitationally pull the given galaxy such as *Milky Way*, to make it rotate. The masses of the galaxies and their distances from their neighboring galaxies are known. Calculations, based on Newton's theory of universal gravitation (as a first approximation to general relativity) then showed that this was inadequate to answer the question about the cause of the rotation of the galaxies. It was then postulated that there must be unseen matter (the 'dark matter') that permeates the entire universe that is responsible through its gravitational coupling with the galaxies for their rotations.

Candidates for this 'dark matter' could be a dense sea of neutrinos or a dense sea of particle-antiparticle pairs. Each of these models of dark matter is charge-independent and exerts gravitational forces on other matter.

Cosmology: The Physics of the Universe

In the first lecture, we discussed Hubble's discovery of the expansion of the universe. That is, each galaxy is moving away from its neighboring galaxies. This is meant in the sense that the density of matter of the universe is decreasing in time, at each spatial point. (It does not mean that the universe is expanding into empty space — there is no empty space; the universe is all that there is!)

If we should extrapolate backward in time, the density of the universe would become steadily greater at earlier times. We would finally reach a state of maximum density and instability of the matter of the universe. At that time, the matter of the universe exploded — this is called the 'big bang.' It initiated an expansion phase of the universe as a whole. Hubble's explicit discovery was that there is a linear relation between the speed of any galaxy, v, relative to another galaxy in this expansion phase and their mutual separation, r, as follows: $v = Hr$. From the measure of Hubble's constant, H, the astronomers determine the approximate time of the 'big bang' — it was about 15 billion year ago.

The question then arises: How did the matter of the universe get into this state of maximum density and instability in the first place? The answer that this is the moment when God created the universe is not acceptable in the context of science! It is a religious answer to a scientific question — thus it is a *non-sequitur* (as we discussed in the first lecture).

The only *scientific answer* that I see to this question (that is consistent with the theory of general relativity) is that before the 'big bang' event, the matter of the universe was imploding — it was contracting with ever increasing matter density. In the theory of general relativity, the terms that play the role of 'force' (the components of the 'affine connection') are not positive-definite. That is, under some physical conditions, such as sufficiently high matter density and relative speeds of matter, the mutual forces of matter can be repulsive, while under other conditions of small matter density and low relative mutual speeds, the 'forces' between matter can be attractive. Thus, at the end of any implosion phase of the universe, an inflection point is reached when the predominantly attractive forces are overcome by predominantly repulsive forces — leading to the 'big bang.' This change then results in the onset of the expansion phase (as we now witness) until the next inflection point, when the predominantly repulsive forces will be overcome by predominantly attractive forces, changing the expansion to a contraction, once again. This goes on indefinitely into the past and into the future. It is the 'oscillating universe cosmology.'

This cosmological model is in competition with the 'single big bang' cosmology. In the latter, there is a singular point in time when the universe started its expansion, and then continues indefinitely into the future. The matter of the universe then becomes continually less dense, transforming to a homogeneous cosmic dust.

With this scenario, the human race seems to be an accident in the cosmic scheme of things! It is formed after the unique, initial big bang, when the matter of the universe cools down sufficiently to allow the formation of stars and planets and the life forms on some of them. But this is not long-lived, in comparison with the age of the universe! For, as the cooling continues, the stars, including our Sun, use up their nuclear fuel and all life forms freeze out of existence. The human race then disintegrates and joins the rest of the cosmic dust of the ever-expanding universe. [*This seems to me to be a very boring scenario*!]

On the other hand, a (*to me, less boring*) cosmological scenario follows from the 'oscillating universe cosmology.' After the expanding universe reaches the inflection point that changes it to a contracting universe (in each of its cycles), the matter of the universe warms up, and eventually the human race (of this cycle) is vaporized in extreme heat. After the next inflection point, the contraction changes to an expansion, once again. The matter of the universe then starts to cool down and eventually stars and planets are formed again and on some of the planets the conditions are right for the formation of life forms and a human race. Thus, in this scenario, the existence of the human race is not an accident! In this sense, the human race is as old and as ordered as the universe itself, as it continues to expand and contract in its infinite numbers of cycles, when one human race is replaced with another in each of the cycles of the universe.

There is another 'oscillating universe cosmology' under consideration by physicists at the present time. It is based on the idea of 'strings,' and their two-dimensional version, called 'branes.' (The strings are 10-dimensional entities that are said to make up the matter of the universe). The 'branes,' as different worlds, collide at the beginnings of each of the cycles.[66]

The string theory is a generalization of quantum field theory, as we discussed in a previous lecture; thus, the latter oscillating universe cosmology is rooted in the bases of the quantum theory.

A problem here is that the quantum field theory is based on linearity, compatible with the expression of a probability calculus and the description of *an open system*. But the universe as a whole is *a closed system*! The theory of general relativity applies to a closed system and is, rather, a fundamentally nonlinear theory, not compatible with a probability calculus. If one of these cosmologies is true to nature, the other must be excluded as a false theory. Indeed, it can be argued that the quantum and relativity theories, applied to any domain, are incompatible, both conceptually and mathematically. (*The details of these differences will be discussed in the next (and final) lecture of this series.*)

The Early Friedman Model[67]

In the 1920s, the mathematician/cosmologist A. Friedman (1888–1925) derived a solution of Einstein's field equations in general relativity that led to a derivation of the Hubble law. However, in my view, it was based on some false premises, even though it led to the empirically correct expansion law of Hubble.

Friedman's solution is based on the premise that there is an absolute (cosmological) time. Its origin — the time of the 'big bang' — is said to be the temporal (absolute) beginning of the universe. But this idea is in contradiction with the theory of relativity wherein time is only a relative measure, dependent on the reference frame in which a law of nature is represented. In relativity theory there is no absolute frame of reference, even that of the universe as a whole!

The derivation of the Friedman solution also depends on two other (empirically) false premises: (1) the matter distribution of the universe is homogeneous and (2) the matter distribution of the universe is isotropic. The first premise, which says that the density of matter at any place in the universe is the same as any other place, is clearly false, as any observation of the night sky would reveal, even with the naked eye! The argument against this is *the speculation* that the observed nonhomogeneity is only a local fluctuation, but that, overall, on the average the matter of the universe is indeed homogeneous. But this is false; we know now, with the high resolution instruments, such as the Hubble telescope, that the galaxies cluster in certain places of the universe and that there are many vacant places in the universe as a whole where there are no stars to be seen! It is just as empirically false to say that the matter of the universe is isotropically distributed, i.e. that the distribution of stars is the same as observed from any angular orientation.

Thus, even though the prediction of the Hubble law from the Friedman solution is empirically correct, the derivation of the solution itself is not valid, from an empirical view and from its incompatibility with the basis of the theory of relativity.

The Hubble Law

While the Hubble law, $v = Hr$, is empirically correct, it is not a covariant expression; i.e. its transformation to other frames of reference would change its form. We must then conclude that, according to the theory of general relativity, this law is a valid *approximation* for a truly covariant law of the dynamics of the matter of the universe.

The Beginning of the Universe

One might ask this question: If the last 'big bang' was only the beginning of this particular cycle of an oscillating universe, then *when* did all of the cycles begin? In Biblical terms, when was the '*beraisheet*' — the beginning of the universe? This is not a scientific question! It is a religious question that can only be answered in terms of one's faith, as a *religious truth.*

The answer cannot be in terms of our human understanding of 'time.' If we believe that God created the universe, *ab initio*, it is illogical to specify *when* this happened, in human terms, be it 5767 years ago, 15 billion years ago, or any time in the infinite past! Time is undefined in the context of this question!

Olbers' Paradox

A problem in astronomy, yet unanswered satisfactorily in the established physics circles, after a very long time, is the following: If the (relatively) infinite numbers of stars of the universe are continually emitting light (and other forms of radiation), why is the night sky dark? Why is it that this infinite amount of emitted light does not fill the universe?

One answer that has been given (by most astronomers) is that the frequencies of light emitted, say in the visible spectrum, are decreased because of the expansion of the universe — the emitter of this light is moving away from us (the absorbers) and its radiation is thereby red-shifted, according to the Doppler effect. With this decrease in the frequency of a given light particle — a photon — its energy (according to the Planck rule, $E = hf$) is thereby decreased, to the point where it becomes zero with respect to our observations.

I do not believe that this is a valid answer to the question. This is because what we 'see' due to the relative motion of the emitters of light, on the one hand, and the *intrinsic energy* of this light, on the other hand, are two different things! The latter is the *proper energy* of the emitted light from the stars — it is not affected by our observations of it! The conclusion would still have to be reached, with this model of light as a collection of freely moving photons, that the total energy of the emitted radiation from the stars of the universe must be approaching infinity! Yet, the night sky is dark! Why is this so? This is Olbers' paradox.

I believe that the answer to Olbers' paradox comes from Faraday's interpretation of light. It is that light is the coupling between emitter and absorber, that are electrically charged matter, to effect their interaction. But with this view, light is not a 'thing in itself' (such a collection of freely moving 'photons'). With Faraday's (and later on, Planck's) interpretation, and the discoveries about light from the theory of relativity, the emitted light only reaches an absorber (say, our eyes or telescopes) that is at a 'timelike' distance from it, i.e. $R < ct$. But most of the stars of the universe are at 'spacelike' distances from us, where $R > ct$. That is, we are too far away from these stars of the universe to absorb their emitted light (and therefore to interact with them) in the time t. In this case, the night sky must be dark to us! This is my answer to Olbers' paradox. It is based on the interpretation of light as not a thing in itself, but only a coupling between interacting matter. And it is based on the theory of relativity.

A Spiral Universe[68]

It is usually assumed in problems of cosmology (as in Friedman's model) that the distribution of matter in the universe is isotropic, as is the expansion of the universe. In my research program, in general relativity, I have found that there are solutions describing an oscillating universe with a spiral, rather an isotropic configuration. Furthermore, in a first approximation to this covariant theory of the universe as a whole, when the relative speeds are small compared with the speed of light and where interstellar distances are small compared with the distance from the observer to the horizon of the universe, the Hubble law follows, as an approximation for the (covariantly described) dynamics of the matter of the universe.

One of the mathematical features in this analysis in general relativity is the replacement of the 10-component metric tensor $g_{\mu\nu}$ of Einstein's formulation with the 16-component quaternion metrical field, q_μ. [*This is a natural consequence of dropping the reflection symmetry in space and time of Einstein's tensor formulation, maintaining only the symmetry with respect to continuous transformations, as required by the principle of relativity — the underlying premise of the theory of relativity.*] The latter is a four-vector field, in which each of the four vector components is quaternion valued, (i.e. each has four components) rather than being real number valued. Thus, the metrical field q_μ has $4 \times 4 = 16$ independent components. Such a formulation then led to a unified field theory, where (in iterated form) 10 of the components relate to the gravitational field and 6 components relate to the electromagnetic field.

A prediction of this formulation of general relativity is that the natural geodesic in spacetime (the trajectory of a test body) is not only in regard to its *natural translation* in space, but also in regard to its *natural rotation* in space. This theory predicts, in principle, that there is a natural rotation of the galaxies (as observed) as well as the spiral configuration of the galaxies (also an observed fact), and a spiral configuration of the universe as a whole. One further prediction is that the plane of polarization of cosmic radiation must rotate as it propagates throughout interstellar space. This effect has been seen in recent astronomical observations.[69]

The Separation of Matter and Antimatter in the Universe[70]

A problem in particle physics and cosmology is the following: Why is there a predominance of matter in our domain of the universe (electrons, protons, ...) rather than antimatter (positrons, antiprotons, ...)?

In my research program, there is a fundamental background of bound particle–antiparticle pairs, each in a particular state, (derived mathematically) that is their ground state, of null energy, momentum and angular momentum. The scenario for the separation of matter from antimatter, at the initial stages of each cycle of the expansion phase of the universe, is the following: At the inflection point, when the contraction phase changes to an expansion phase and after some cooling has taken place, the gravitational field of the matter of the universe delivers about 1 MeV units of energy to each of a small portion of the electron–positron pairs, to dissociate them, *out of many more pairs that do not receive this energy*. Similarly, about 2 GeV units of energy are delivered to some of the proton–antiproton pairs in their ground states. (The remaining electron and proton pairs, in their ground states of null energy, momentum and angular momentum, are the *dark matter* that we discussed previously.) The released particles and antiparticles are then in a rotational motion of the spiraling universe, along with the (many more) neutral pairs of the universe as a whole. The rotating particles and antiparticles, in a plane perpendicular to the axis of rotation of the universe, being oppositely charged, create magnetic fields parallel and antiparallel to the axis of rotation of the spiraling universe. Thus, there is a competition between the gravitational field of the matter of the universe, inducing particles and antiparticles to move in a single rotational direction of the spiraling universe, and the magnetic fields separating the directions of motion of the particles and antiparticles. More particles than antiparticles will move in one direction and more antiparticles than particles will move in the opposite direction. Thus, matter and antimatter become separated in the early expansion stages of the cycles of the universe.

It is possible, then, that in a different region of our universe, there is mostly antimatter, where, for example, (negatively charged) antiprotons and (positively charged) positrons bind to form anti-Hydrogen, where nuclei of two antineutrons bound (with the nuclear force) to two (negatively charged) antiprotons bind electrically to two (positively charged) positrons to form an atom of anti-Helium, and where there are complex structures, such as planets, stars, galaxies, ... *and human beings*, that are composed of antimatter. This is a speculation that may be empirically testable in future studies.

In the next (and final) lecture, we will discuss the foundational (and irresolvable) differences between the bases of the quantum and relativity theories, both claiming to explain the truths about the nature of matter.

Lecture X

CONFLICTS IN THE FOUNDATIONS OF THE QUANTUM AND RELATIVITY THEORIES

During this course of lectures on concepts of modern physics, we have emphasized the bases of the two major developments in the 20th century: the quantum and relativity theories. Each claims to be an explanation for the nature of matter. The trouble encountered with the appearance of two simultaneous theories, rather than one at a time, is that each requires an incorporation of the other for its completion.[71] On the other hand, each of these theories is incompatible with the other, both conceptually and mathematically, as we will discuss in this lecture. My resolution of this dichotomy is to totally abandon the basis of one of these theories as a conceptual truth, and to then generalize the other so as to recover the empirically correct features of the abandoned theory, as an approximation for some aspect of the accepted theory.

In this (the last) lecture of the series, I will try to spell out some of the main differences in the foundational concepts of the quantum and relativity theories. If, as I claim, these are irresolvable, it signifies that there can be no union of the two theories, either mathematically or conceptually. Thus there can be no quantum theory of gravity, or any sort of relativistic quantum field theory. But in principle the mathematical features of the quantum theory that match the empirical facts could be an approximation for predictions of the theory of relativity, or vice versa. It should then be up to the physicists who study this dichotomy to decide for themselves which of these theories to abandon and which to keep, for the direction of science toward its ultimate goal — *the truth* about the natural world.

The Principle of Complementarity Versus the Principle of Relativity

A primary basis of the quantum theory is *Bohr's principle of complementarity.*[72] This is a philosophy of *pluralism.* That is to say, this is an assertion (not only in physics) that there are many logically dichotomous truths; they are all acceptable so long as they are established at different times. In the physics of elementary matter it is the view of wave-particle dualism. Matter can be formed of discrete particles *and* it can be formed of continuous waves, so long as these qualities are measured at different times, under different sorts of experimental conditions. In this view, the truth about some aspect of a material system (be it an electron or a galaxy of stars) is then dependent on a human being's perceptions of it.

On the other hand, the theory of relativity is based on a philosophy of *monism* — this is implied by *Einstein's principle of relativity.* It is the idea that the laws of nature must be fully *objective.* That is to say, the laws of nature, *per se,* are independent of the perspectives from which they are determined. If an electron is a discrete particle then it cannot be a continuous wave, and vice versa. Thus, according the monistic approach of the theory of relativity, even though an electron may *appear* to our apparatuses (our human senses or sophisticated instruments) to be a discrete particle, it cannot be so if experimentation reveals it also to be a continuous wave. Experimentation must then determine which of these qualities is the correct one, objectively — the particle or the wave — independent of the observer!

In the field concept implied by the theory of relativity, there are no separate, singular particles of matter. Instead, there are (an infinite number of) distinguishable modes of a single entity — the universe. This is a *holistic* theory of matter. In this view, all objects are manifestations of a single underlying order. Such a single underlying order of matter is then expressed with fully objective laws of nature. Thus, there cannot be any conceptual lines of demarcation between one set of laws, to underlie one set of physical phenomena, and another set of laws, to underlie a different set of physical phenomena. Thus, Bohr's principle of complementarity is automatically rejected by the holistic approach of Einstein's theory of relativity. Bohr's epistemological view of pluralism is that of *positivism* while Einstein's epistemological view of monism is that of *realism*.

Atomism Versus Continuity

A major difference between the quantum and relativity theories regarding the fundamental nature of matter has also been debated since ancient Greece. It is the question about whether matter is fundamentally atomistic or a continuum. In ancient Greece one identifies these opposite views with Democritus, for atomism, and Parmenides, for continuity.

According to the quantum theory, matter is atomistic, composed of singular, elementary particles that may (or may not) interact with each other. This is different than the classical atomistic theories in that the physical properties of the elements of matter, here, are not predetermined — they depend on the conditions of measurement. But at the root of matter are still the elementary particles — the electron, the proton, etc. [*In the later stages of physics, at the present time, it has been speculated that the units of matter are 10-dimensional strings. This was initially proposed to overcome the problem of the infinities that appear in 'relativistic quantum field theory' of the three-dimensional discrete (point) elementary particles. The theory of strings has also been generalized to membranes — called 'branes.' Unfortunately, after many years of theoretical studies, the 'string theory' has not yet yielded a mathematically consistent (finite) quantum field theory of matter, which was its original purpose, nor has it successfully predicted any observable facts.*]

In contrast with the atomistic theory of the quantum view, the theory of relativity, when pursued to its logical extreme, implies, fundamentally, a continuum field theory of matter. What appear as the elementary particles of matter are instead the distinguishable manifestations of a single continuum that is the universe. The model of matter is then a *holistic* ontology. There are no separate 'things' — electrons, human beings, planets, stars, galaxies, ... — these are all correlated modes of a single universe. In philosophy, this is in the sense of Spinoza.[73]

The analogy that I have used is that, in the holistic, continuum view, the 'particle of matter' is like the ripple of a pond. The ripple is not an individual, separable thing. It cannot be removed from the pond, and its properties then determined — its weight, color, size, etc. The ripple is a mode of the entire, continuous pond. If the pond had infinite boundaries, then the ripple would be everywhere, all at once!

[*If you will indulge with this physicist in putting on my 'philosophical hat,' I would say that if human beings could see themselves in this holistic way, forgetting the ego-building speculations about themselves, leading to their seeking power over others and unobstructed personal achievements, it may become a better world, without wars or prejudices, leading to the harming of other human beings. If indeed physics should move in this direction that is indicated by the continuous, holistic field concept, then perhaps the idea would creep into the human culture and improve our lives.*]

On Epistemology – Logical Positivism Versus Abstract Realism

The theory of knowledge (epistemology), according to the quantum theory, has been indicated earlier to be that of *logical positivism.* This is an approach to knowledge based on the 'principle of verifiability.' It is the assertion that the only meaningful statements in science, when expressed in a mathematical or verbal language, must be empirically verifiable. This was Heisenberg's explicit statement, when he embarked on the development of the quantum theory.[74] In Bohr's philosophy this is a general statement about knowledge, even beyond the domain of atomic and elementary particle physics. In physics, the concept of wave-particle dualism supports this view.

On the other hand, the theory of relativity is based on an epistemology of *abstract realism*. This is an approach to knowledge that assumes that there is a real world 'out there,' independent of whether or not we are there to observe its physical properties.

Of course, physics is an empirical, science — its truths depend on confirmation with the observational facts. But this does not mean that there are not other features of a true theory of matter — its explanation — that are needed logically, but are not in themselves observables (this is the reason for the adjective 'abstract'). It is indeed our rational analyses of a theory in science and its logical conclusions that tell us what to believe are its (scientific) truths. In philosophy, this is in the sense of Plato.

An example of a physical theory that is understood in terms of *abstract realism* is the Faraday–Maxwell field theory of electromagnetism. The solutions of the Maxwell field equations are not directly observable. It is only after inserting these field variable solutions into the forms for the energy, momentum, angular momentum and force on charged matter, that we can make predictions of the effects of the electromagnetic variables. If theory and experiment then agree, we are permitted to say that there is scientific truth in the theory that we started from.

Subjectivity Versus Objectivity

This difference between the bases of the relativity and quantum theories has to do with the subjective view of the Copenhagen School on the *fundamental* incorporation of the macroscopic measuring apparatus with the measured microscopic matter, in contrast with the fully objective approach of the theory of relativity, regarding the nature of the observed matter.

The claim of the Copenhagen School is that there is no causal relation between the act of the apparatus in measuring the properties of micromatter and the determined properties of this matter. Bohr said the following: "The very existence of the quantum of action entails the necessity of a final renunciation of the classical idea of causality and a radical revision of our attitude toward the problem of physical reality."[75]

Bohr then assumed that the physical description of the measuring apparatus must be governed by the classical set of rules for macromatter while the physics of the observed matter is governed by the rules of quantum mechanics. Thus we see, at the outset, that according to Bohr's philosophy, one must start with the (acausally) coupled system of *measurer* and *measured*, distinguishing each by the location of a line of demarcation between them.

According to the quantum theory, there is no strict rule about the location of the line of demarcation between the *measurer* and the *measured*, except that the classical action of the *measurer* must be large compared with the quantum of action that is Planck's constant h. Still, one may move this line, arbitrarily, while keeping the required inequality between the mechanical actions of the classical and microscopic domains. In this case, the predictions about the *measured* would correspondingly change. The properties (of an electron, a hydrogen molecule, etc.) are then in part *subjective*. That is to say, the physical properties of the observed microscopic matter depend on the nature of the *measurer* — the classical measuring apparatus.

In contrast, the theory of relativity requires at the outset that the laws of nature must be independent of the reference frame in which they are represented, be this the reference frame of the *observer* or that of the *observed*. Thus, there is no difference between the variables of the *observer* and those of the *observed*, as there is in the quantum theory. This is because relativity theory represents a *closed system* at the outset. The implication here is that the variables of the *observer* and those of the *observed* must be interchangeable, without altering the fully objective description of their interaction.

For example, consider an electromagnetic interaction: an emitter sends a signal (a 'photon') to an absorber, in a finite time. The implication is that the emitter will not send the signal unless there will be an absorber present at the later time to absorb it. The only purpose of the signal is to act as a connecting link between the emitter and the absorber to effect their mutual interaction. (This is called "delayed action at a distance."[76]) The symmetry of the theory of relativity then implies that to be fully objective, the reference frame of the emitter must be interchangeable with the reference frame of the absorber, without affecting the description of their mutual interaction. Thus, with this interchange, the emitter becomes the absorber and the absorber becomes the emitter. It implies that for every signal emitted to a distant absorber, there must be another signal emitted simultaneously by the absorber to the emitter! (This is an implication of relativity theory that has yet to be tested empirically.)

[*I had the opportunity to discuss this concept of 'delayed action at a distance' with Richard Feynman, in the 1950s. His PhD thesis was based on this topic. He (and his advisor, John Wheeler) suggested a system of matter that is all particles (each in their own spacetime) and no fields, interacting with delayed action at a distance. It led to difficulties that entailed a requirement of the addition of a many-particle system and the need for a statistical description for the way of representing a single emitter and absorber. Feynman felt that, nevertheless, the delayed action at a distance concept should be salvageable. It was my suggestion that instead of all particles and no fields, of an open system, as he and Wheeler proposed, the symmetry of relativity theory requires delayed action at a distance but for a closed system of all fields (each in a common spacetime) and no particles! But this approach would then lead to a nonlinear formal expression of the material system in terms of coupled nonsingular fields that is not compatible with the quantum theory — thus largely unacceptable to the adherents of the quantum theory!*]

On Quantum Electrodynamics

It is interesting that the elementarity of the interaction in relativity theory, or the elementarity of measurement in the quantum theory, imply that there must be a fundamental *unbreakable* triad — *emitter–signal–absorber* — to start with. In electrodynamics the signal is a 'photon.' This signal has no nonrelativistic limit, (as Einstein originally discovered), while the material emitter and absorber components of the unbreakable triad do have nonrelativistic limits in their *descriptions.*

Thus it is essential that the *unbreakable* triad, *emitter–signal–absorber,* be represented relativistically at the outset. Once this is accomplished, one may then take the limit of the emitter and the absorber components of the triad in a nonrelativistic approximation, to achieve their description in nonrelativistic quantum mechanics. But when this attempt was made, at the early stages of quantum electrodynamics, to combine the matter and the radiation it emits and absorbs in this way, it was found that the solutions were described by infinite divergent series. This is the theory called "quantum electrodynamics." Later on, renormalization methods were discovered that allowed one to subtract off the infinities of the divergent series solutions. This resulted in new predictions in the atomic domain, in agreement with the empirical facts, that were not predicted at all with normal quantum mechanics. (Such were the Lamb shift in hydrogenic atoms and the anomalous magnetic moment of the electron.) While this was taken to be a success of the quantum theory, it is still not satisfactory because the renormalization procedures are not mathematically consistent. That is, while some predictions are correct, by changing the method of subtracting the infinities from the divergent series solutions, one may predict any other numbers for the same physical effects! This violates the scientific requirement that there is a unique prediction for any given experimental fact.

Thus it has been my contention, as well as some others in the field (such as one of the original founders of quantum field theory,

Paul Dirac[77]) that quantum electrodynamics is not in a satisfactory state as a bona fide theory. My own research program has been to show that the quantum theory formalism may be a linear approximation for a theory of the inertia of matter, based on the theory of general relativity. This theory has thus far given all of the results given by ordinary quantum mechanics, as well as results given by quantum electrodynamics, (such as the Lamb shift), but in a mathematically consistent way.[78]

Indeterminism Versus Determinism

The classical theories in physics, as well as Einstein's theory of relativity, are *deterministic*. This is in the sense that the entire physical system is *predetermined*. In Newtonian physics, it is the (absolute) time parameter that does the ordering. That is, given the state of a physical (generally many-body) system, at some initial time, the entire future of the physical system is predetermined. (This is called the 'Cauchy problem.') In relativity theory, where the time parameter is mixed in with the space parameters in an objective spacetime, the ordering parameter for the future of a physical system is not necessarily the time measure alone. It is any parameters that provide a logical ordering of the material system.

On the other hand, as we discussed earlier, the material system, according to the quantum theory, is not predetermined. The physical properties of micromatter are determined only after a measurement has been carried out. It is in this sense that the quantum theory is *nondeterministic*. This is an important, irresolvable difference from the *determinism* of the theory of relativity.

Summing up, we have discussed five of the fundamental conceptual differences between the quantum and relativity theories: (1) the principle of complementarity (pluralism) versus the principle of relativity (monism), (2) atomism versus continuity, (3) logical positivism versus realism, (4) subjectivity versus objectivity and (5) indeterminism versus determinism.

From the mathematical side, the quantum theory is in principle a linear theory because its solutions relate to the probabilities of measurement, which in turn must necessarily be expressed with a linear formalism (i.e. the sum of any number of solutions is another solution of the underlying equations). The theory of relativity, on the other hand, is necessarily a nonlinear theory. This is because (1) the fundamental equations of any material system represent a *closed system* and (2) it generally represents a system of matter with a curved spacetime.

The foregoing are *fundamental differences* in the two approaches to an explanation of the laws of matter. Thus it must be concluded that the theory of relativity and the quantum theory are conceptually and mathematically incompatible. Aside from approximations that may be used for the purpose of calculations, (e.g. a linear approximation for a nonlinear mathematical formalism) if one of these theories is true to nature the other must be false. The student of physics must then choose which of these two theories is the one that is more scientifically valid to pursue in the further development of physics in the 21st century. My own intuition and extensive research results tells me that it is the theory of relativity where the truth lies. Time will tell!

REFERENCES AND NOTES

1. I discussed this dichotomy in: M. Sachs, *Einstein Versus Bohr* (Open Court, 1988), Chapter 10.
2. T. S. Kuhn, *The Structure of Scientific Revolutions* (Chicago, 1970), *second edition*. See also: K. R. Popper, *The Logic of Scientific Discovery* (Harper, 1959).
3. Aristotle's philosophy is expounded in: R. McKeon, editor, *The Basic Works of Aristotle* (Random House, 1941), Metaphysics (trans. W. D. Ross).
4. For a concise outline of Einstein's views of science, see: A. Einstein, *Ideas and Opinions* (Crown Publishers, 1959), Part V, Contributions to Science.
5. See: T. V. Smith, *editor, From Thales to Plato* (Chicago, 1956), Chapter V.
6. E. P. Hubble, *The Realm of Nebulae* (Yale, 1936), Chapter 1.
7. Galileo Galilei, *Dialogue Concerning Two Chief World Systems*, S. Drake, *translator* (California, 1970), Second edition, p. 101.
8. This idea is discussed in: S. W. Hawking, *A Brief History of Time* (Bantam, 1990), Chapter 10.
9. Galileo's principle of inertia is explicated in: E. Mach, *The Science of Mechanics* (Open Court, 1960), p. 168.
10. Galileo's discussion of projectile motion is in: Galileo Galilei, *Two New Sciences*, S. Drake, *translator*, (Wisconsin, 1974).

11. Ref. 8, p. 124.
12. Ref. 7.
13. For discussion of Galileo's law of gravity, see: M. Sachs, *Relativity in Our Time* (Taylor and Francis, 1993), p. 55.
14. E. S. Haldane and G. T. Ross, *translators, The Philosophical Works of Descartes* (Cambridge, 1934). On Descartes' debate with Spinoza, see: B. Spinoza, *The Principles of Descartes' Philosophy* (Open Court, 1974).
15. B. Spinoza, *The Ethics, Correspondence, Works of Spinoza,* edited by R. Elwes (Dover, 1955).
16. Newton's major published works are: I. Newton, *Principia, Vol. I,* The Motion of Bodies; *Principia, Vol. II,* The System of the World (California, 1934), F. Cajori, *translator. Optiks* (Dover, 1952).
17. The nonlinearity of the description of a closed system is exemplified in Newton's third law of motion. Consider a body X obeying Newton's second law: $F_X = m_X a_X$. The force F_X is exerted on X by the body Y, a distance away from it. The body Y, in turn, is acted upon by X, according to Newton's third law, obeying the force law, $F_Y = -F_X = m_Y a_Y$. The subsequent motion of Y with acceleration a_Y then changes its initial position when it acts upon X, causing X, in turn, to change its motion because of the recoil and the subsequent changed position of Y. The force F_X then depends on a_X itself, by virtue of the mediating influence of body Y. Thus the variables of motion of X depend on its own variables of motion — the equation in X is then nonlinear in terms of the acceleration of X. Similarly the motion of Y depends on its own motion and its law of motion is similarly nonlinear. I do not believe that Newton ever noted this mathematical change from linearity to nonlinearity induced by his third law of motion, implying a closed system at the outset.
18. As we will discuss in the next lecture, these modes are similar to Spinoza's view of the universe, not a collection of separable things, but rather holistically, analogous to the inseparable ripples of a pond.

19. Boltzmann discovered the kinetic theory of gases. This subject is developed in: E. H. Kennard, *Kinetic Theory of Gases* (McGraw-Hill, 1938).
20. H. Reichenbach, *The Direction of Time* (California, 1956), Chapter 3.
21. Bertrand Russell refuted the stand of naive realism with this argument: "The observer, when he seems to himself to be observing a stone, is really, if physics (physiology) is to be believed, observing the effect of the stone upon himself. Thus, naïve realism, if true, is false; therefore it is false" (quoted in A. Einstein, "Remarks on Bertrand Russell's Theory of Knowledge", in P. A. Schilpp, *editor, The Philosophy of Bertrand Russell* (Open Court, 1944), p. 283.
22. In Ref. 9, p. 342, Mach said: "The history of science teaches that the subjective, scientific philosophies of individuals are constantly being corrected and obscured, and in the philosophy of constructive image of the universe which humanity gradually adopts, only the very strongest features of the thoughts of the greatest men are, after some lapse of time, recognizable."
23. The Mach principle is discussed in: M. Sachs, The Mach Principle and the Origin of Inertia from General Relativity", M. Sachs and A. R. Roy, *editors, Mach's Principle and the Origin of Inertia* (Apeiron, 2003).
24. M. Sachs, *The Field Concept in Contemporary Science* (Thomas, 1973).
25. J. C. Maxwell, *A Treatise on Electricity and Magnetism*, Vol. 1 (Dover, 1954), *third edition*; Vol. 2, (Dover, 1954), *third edition.*
26. A. Michelson and E. Morley, *Silliman Journal* **34**, 333, 427 (1887).
27. M. Planck and M. Masius, *The Theory of Radiation* (McGraw-Hill, 1914).
28. J. J. Thomson, "Cathode Rays", *Philosophical Magazine* **44**, 293 (1897).
29. R. A. Millikan, *The Electron* (Chicago, 1924), *second edition.*
30. A. Einstein, "The Photoelectric Effect", in M. H. Shamos, *editor, Great Experiments in Physics* (Dover, 1959), p. 232.

31. A. Compton, "The Compton Effect", in Shamos, *ibid.*, p. 348.
32. H. Becquerel, "Natural Radioactivity", in Shamos, *ibid.*, p. 210.
33. J. Chadwick, "The Neutron", in Shamos, *ibid.*, p. 266.
34. E. Rutherford, *Philosophical Magazine* **26**, 1 (1913).
35. N. Bohr, "The Hydrogen Atom", in Shamos, *ibid.*, p. 329.
36. C. J. Davisson and L. H. Germer, "Diffraction of Electrons by a Crystal of Nickel", *Physical Review* **30**, 705 (1927); G. P. Thomson, "Experiments on the Diffraction of Cathode Rays", *Proc. Roy. Soc. (London)* **A117**, 600 (1928).
37. Further comments on wave-particle dualism are in: M. Sachs, *Einstein Versus Bohr* (Open Court, 1988); M. Sachs, "On Wave-Particle Dualism", *Annales de la Fondation Louis de Broglie* **1**, 129 (1976).
38. For further discussion see: M. Sachs, *Einstein versus Bohr, ibid.*, p. 103.
39. M. Sachs, *ibid.*, p. 107.
40. For an English translation of Heisenberg's paper, see: W. Heisenberg, "Quantum Theoretical Re-Interpretation of Kinematic and Mechanical Relations" van der Waerden, *editor, Sources of Quantum Mechanics* (Dover, 1968), p. 261.
41. M. Sachs, *ibid.*, pp. 121–136.
42. C. Lanczos, (in German), English translation of the title: "On a Field Theoretical Representation of the New Quantum Mechanics", *Zeits. fur Physik* **35**, 812 (1926). Also see: M. Sachs, "Cornelius Lanczos — Discoveries in the Quantum and General Relativity Theories", *Annales de la Fondation Louis de Broglie* **27**, 85 (2002).
43. A rigorous derivation of the Heisenberg uncertainty relations is in: M. Born, *Atomic Physics* (Hafner, 1946), Appendix XXII.
44. For a history of the principle of complementarity, see: G. Holton, *Thematic Origins of Scientific Thought* (Harvard, 1973). Also see: N. Bohr, *Atomic Physics and Human Knowledge* (Wiley, 1958).
45. M. Sachs, *Einstein Versus Bohr, ibid.*, pp. 139–143.
46. A. Einstein, B. Podolsky and N. Rosen, "Can Quantum Mechanical Description of Physical Reality be Considered Complete?" *Physical Review* **47**, 777 (1935).

47. N. Bohr, "Can Quantum Mechanical Description of Physical Reality be Considered Complete? *Physical Review* **48**, 696 (1935).
48. L. de Broglie, *The Current Interpretation of Wave Mechanics* (Elsevier, 1964), Chapters 4, 6.
49. D. Bohm, "A Suggested Interpretation of the Quantum Theory in Terms of Hidden Variables", *Physical Review* **85**, 166, 180 (1952).
50. J. von Neumann, *Mathematical Foundations of Quantum Mechanics* (Princeton, 1955).
51. A general discussion is given in: M. Sachs, *Relativity in Our Time* (Taylor and Francis, 1993).
52. M. Sachs, "A Resolution of the Clock Paradox", *Physics Today* **24**, 23 (1971). A debate on this question is in the 'Letters section' of *Physics Today* **25**, 9 (1972).
53. M. Harada and M. Sachs, "Re-Interpretation of the Fitzgerald-Lorentz Contraction", *Physics Essays* **11**, 521 (1998).
54. To eliminate this paradox, some physicists appeal to the many-universe idea. What they say is that when the surgeon meets his (much younger!) father, and destroys his own possibility of existing by making his father sterile before the surgeon was born, this is in one universe, but in a different (parallel) universe he still exists. For this argument, see, for example, S. W. Hawking, *The Universe in a Nutshell* (Bantam, 2001), p. 135. *The student should make up his own mind about the meaningfulness of this resolution!*
55. K. Gödel, "An Example of a New Type of Cosmological Solution of Einstein's Field Equations", *Reviews of Modern Physics* **21**, 447 (1949).
56. A. Einstein, "Does the Inertia of a Body Depend on its Energy Content?" in: A. Einstein, H. A. Lorentz, H. Weyl and H. Minkowski, *The Principle of Relativity* (Dover, 1929). Also see: M. Sachs, "The Meaning of $E = mc^2$", *International Journal of Theoretical Physics* **8**, 377 (1973).
57. A. Einstein, *The Meaning of Relativity: Relativity Theory of the Nonsymmetric Field* (Princeton, 1956), *fifth edition*.

58. M. Sachs, "On the Logical, Status of Equivalence Principles in General Relativity Theory", *British Journal for the Philosophy of Science* **27**, 225 (1976).
59. Predictions of the three crucial tests of general relativity are outlined in: R. J. Adler, M. J. Bazin and M. Schiffer, *Introduction to General Relativity* (McGraw-Hill, 1975), *second edition*.
60. R. V. Pound and G. A. Rebka, "Gravitational Redshift in Nuclear Resonance", *Physical Review Letters* **3**, 439 (1959); "R. V. Pound and J. L. Snider, "Effect of Gravity on Nuclear Resonance", *Physical Review Letters* **13**, 539 (1964).
61. E. Schrödinger, *Space-Time Structure* (Cambridge, 1954).
62. An account of this view of a "black hole" is in: S. W. Hawking and W. Israel, editors, *300 Years of Gravitation* (Cambridge, 1987), Chapter 7 (by W. Israel) and Chapter 8 (by R. D. Blandford).
63. Hawking and Israel, *ibid.*, pp. 116–120.
64. M. Sachs, "A Pulsar Model from an Oscillating Black Hole", *Foundations of Physics* **12**, 689 (1982).
65. M. Sachs, "On the Rotations of Galaxies from General Relativity", *Physics Essays* **7**, 490 (1994).
66. For a clear discussion of this cosmological model, see the website of Professor Paul Steinhardt, at: www.physics.princeton.edu/ ~ steinh
67. Adler, Bazin and Schiffer, *ibid.*, p. 435.
68. M. Sachs, *Quantum Mechanics and Gravity* (Springer, 2004), Sec. 8.7.3.
69. B. Nodland and J. P. Ralston, "Indication of Anisotropy in Electromagnetic Propagation over Cosmological Distances", *Physical Review Letters* **78**, 3043 (1997).
70. M. Sachs, "On the Separation of Matter and Antimatter in the Early Universe", *Annales de la Fondation Louis de Broglie* **20**, 323 (1995).
71. M. Sachs, *Einstein Versus Bohr* (Open Court, 1988), Chapter 10.
72. N. Bohr, *Atomic Physics and Human Knowledge* (Wiley, 1958).
73. B. Spinoza, *Ethics*, translated by R. H. Elwes (Dover, 1955).

74. See Ref. 40.
75. See Ref. 47.
76. J. A. Wheeler and R. P. Feynman, "Classical Electrodynamics in Terms of Direct Interparticle Interaction", *Reviews of Modern Physics* **21**, 425 (1949). Also see: M. Sachs, *Einstein Versus Bohr* (Open Court, 1988), p. 227.
77. P. A. M. Dirac, "The Early Years of Relativity", in: G. Holton and Y. Elkana, *editors*, *Albert Einstein: Historical and Cultural Perspectives* (Princeton, 1982), p. 79. In this article, Dirac made the following comment: "It seems clear that the present quantum mechanics is not in its final form. Some further changes will be needed, just about as drastic as the changes made in passing from Bohr's orbit theory to quantum mechanics. Some day, a new quantum mechanics, a relativistic one, will be discovered, in which we will not have these infinities occurring at all. It might very well be that the new quantum mechanics will have determinism in the way that Einstein wanted."
78. The development of this research is outlined in: M. Sachs, *Quantum Mechanics and Gravity* (Springer, 2004).

INDEX